The Team Based Product Development Guidebook

Also available from ASQ Quality Press

Business Process Improvement Toolbox
Bjørn Andersen

Team Fitness: A How-To Manual for Building a Winning Work Team
Meg Hartzler and Jane E. Henry, Ph.D.

Lessons from Team Leaders: A Team Fitness Companion
Jane E. Henry, Ph.D.

Tools for Virtual Teams: A Team Fitness Companion
Jane E. Henry and Meg Hartzler

Tool Navigator: The Master Guide for Teams
Walter J. Michalski with Dana G. King

The Toolbox for the Mind: Finding and Implementing Creative Solutions in the Workplace
D. Keith Denton with contributions from Rebecca A. Denton

Principles and Practices of Organizational Performance Excellence
Thomas J. Cartin

Mapping Work Processes
Diane Galloway

Success through Quality: Support Guide for the Journey to Continuous Improvement
Timothy J. Clark

Quality Problem Solving
Gerald F. Smith

To request a complimentary catalog of ASQ Quality Press publications, call 800-248-1946.

The Team Based Product Development Guidebook

Norman B. Reilly

ASQ Quality Press
Milwaukee, Wisconsin

The Team Based Product Development Guidebook
Norman B. Reilly

Library of Congress Cataloging-in-Publication Data

Reilly, Norman B.
  The team based product development guidebook / Norman B. Reilly.
    p. cm.
  Includes index.
  ISBN 0-87389-451-0
  1. Product mangement.   2. Teams in the workplace.   3. New
products.   I. Title.
  HF5415.153.R45   1999
  658.5'75—dc21   99-24687
                                                                   CIP

© 1999 by ASQ

10 9 8 7 6 5 4 3 2

ISBN 0-87389-451-0

Acquisitions Editor: Ken Zielske
Project Editor: Annemieke Koudstaal
Production Coordinator: Shawn Dohogne

ASQ Mission: The American Society for Quality advances individual and organizational performance excellence worldwide by providing opportunities for learning, quality improvement, and knowledge exchange.

Attention: Bookstores, Wholesalers, Schools and Corporations:
ASQ Quality Press books, videotapes, audiotapes, and software are available at quantity discounts with bulk purchases for business, educational, or instructional use. For information, please contact ASQ Quality Press at 800-248-1946, or write to ASQ Quality Press, P.O. Box 3005, Milwaukee, WI 53201-3005.

To place orders or to request a free copy of the ASQ Quality Press Publications Catalog, including ASQ membership information, call 800-248-1946. Visit our web site at http://www.asq.org.

Printed in the United States of America

∞ Printed on acid-free paper

American Society for Quality

Quality Press
611 East Wisconsin Avenue
Milwaukee, Wisconsin 53202
Call toll free 800-248-1946
http://www.asq.org
http://standardsgroup.asq.org

Many many thanks to Patricia Marion Donnelly

CONTENTS

ACKNOWLEDGMENTS

To all those who have helped me learn how to learn and had the patience to wait.

It is also a distinct pleasure to thank the highly professional staff at the American Society for Quality/Quality Press. In particular, the continued support and sage advice of Ken Zielske, Annemieke Koudstaal, and their editorial staff have immensely contributed to the quality of this book.

Introduction

WHY THIS BOOK?

If you have *anything* to do with product or system development, this book is for you. It provides fundamental guidelines gleaned from real-world experience and considers the complete set of issues that should be addressed at each stage of product development. Completeness is important. The guidelines are derived from experiences of some pretty good organizations and individuals, as well as my own. Every guideline in this book has been hard won. You may find that you already routinely address some of the end-to-end product development issues to be covered. If you are experienced, some of them may even appear obvious to you. But be careful, particularly if you *are* experienced. Respectfully, we need only recall the FAA Advanced Automation System, the Denver Automated Baggage System, The American Airlines Integrated Reservation System, Microsoft recalls of software in China and Mexico, most big-city 911 systems, our repeated attempts at a space station, the Edsel, the Caprice, the Disney cultural gaff outside Paris, the Patriot Missile, the IRS Computer System, and the Hubble Telescope to realize that we can do better. Having made more than a few significant mistakes in my own career, I am also very much in the club. The point is, all of these products were built by very experienced people. The question then becomes: How can we better prepare ourselves to routinely think of everything? This book is designed to answer that question. It is sound and useful for the experienced, as well as a basic resource and guideline for the beginner.

One more thought for the experienced. While you are reading this book, there will be a natural tendency to think, or utter, these famous last words:

"That's a good idea, but it won't work here,"
or
"We're already doing that."

If this happens to you, slow down and think about it very hard.

The purpose of this book is to assist product development personnel in getting to market faster, on schedule and within budget, with quality products that meet true user needs, through the reduction of errors of omission and by maximizing the positive motivation of an organization's most valuable asset—its people.

The methodology to achieve this goal is called team based product development.

WHAT IS TEAM BASED PRODUCT DEVELOPMENT?

Team based product development is a process that utilizes corporate-wide interdisciplinary expertise for (a) initial assessment of product development feasibility, (b) complete planning for product development, and (c) concurrent execution of product development.

By definition, this is not a typical "team" book. The great majority of today's team oriented books deal with the temporary establishment of teams to define and solve specific one-time problems. Such team efforts typically address microprocesses within the organization, and upon achieving process improvements, are disbanded. Product development teams are different. They are created for each product to be developed. They then stay with the processes during and through the completion of each product development phase.

WHAT ARE THE PHASES?

There are three fundamental stages that we all go through in the development of new, or improved, products. Your organization may have additional "gates" or "milestones," even inchstones, but there are only three fundamental steps. Informally, these steps answer the questions: (1) Is this a good idea or is it nuts? (2) If it sounds good, how are we actually going to do this? and (3) If it still looks right, what is the actual process to be followed in developing a product that meets stated requirements, on time, and on schedule? More formally, the first phase investigates the top-level feasibility of product realization and its market viability. If the product idea proves feasible for a given organization, the second step is devoted to construction of a detailed development plan. If the detailed plan is approved, the actual physical product development is then carried out. These phases are referred to as the product development feasibility phase (Phase A), the detailed product development planning phase (Phase B), and the concurrent product development phase (Phase C).

WHY TEAMS?

Teams are important because nobody knows everything. (Not even me.) Well-organized corporate-wide teams can substantially reduce the amount of time devoted to all aspects of the complete product development process. They also significantly enhance the probability that the final product meets intended expectations. Teams can do this because they provide an interdisciplinary knowledge base. The knowledge base includes expertise in marketing, finance, engineering, design, procurement, fabrication, production, assembly, logistics support, integrated costing and scheduling, customer support, suppliers, systems engineering, risk analysis, decision analysis, specialty engineering, quality, testing, users (or faithful user representation), dealers, and anyone else needed to support a complete picture of a successful end-to-end product development process.

In the product development setting, the interdisciplinary team approach substantially contributes to accuracy and completeness, reduces errors of omission, enables the execution of concurrent strategies, reduces surprises, reduces risk, enhances the thoroughness and consistency of the review process, breaks down organizational barriers, saves time through organized and cooperative focus on the right issues in the right sequence, substantially contributes to getting to market faster, and promotes pride of ownership in outcome.

This is proven stuff based on real experiences. It works when the teams work. Make no mistake about it, teams work when individuals understand their roles and are empowered to contribute their expertise. Teams work when individuals have self-esteem, when they have a feeling of importance, when they get up in the morning and want to go to work, and when they can say, "I'm needed here." This single ingredient is common to all successful quality movements. We don't need a lot of current references to corroborate this tenet. It's only about ten thousand years old. But we do need to begin to appreciate and utilize it more.

THREE MORE COMMENTS

1. You don't have to reorganize the whole corporation to implement team based product development. It can be, and has been, done within existing structures.

2. The material in this book is a guide to help you think of everything. It is based on a lot of experience—some good, some not so good. The intent is to be supportive of your ongoing product development efforts.

3. The material presented here is generic to all products—mass produced and one-of-a-kind. It is likely that you are already addressing some of the items to be covered. I would be surprised if you're not. That is not the point. The point is that *every* item should still be considered. It is easier to take something off a list than to consistently think of everything in advance. You decide whether you wish to incorporate some of the ideas or all of them. The more you consider, however, the safer you're going to be. Don't skip anything.

BOOK ORGANIZATION

The book is organized in seven parts. Following Part 1 "Introduction," Part 2 covers the "What" and "Why" of Teams. Valuable and instructive historical anecdotes on how teams have been organized and used in the past are presented. Successful team ingredients with multiple real-life examples from a range of different industries are described. Part 2 concludes with some thoughts of both Fukayama and Socrates on the rewards of building a sense of self-worth among employees.

Part 3 deals with the need for management support and the definition of systems engineering, and presents guidelines on the team leader's mindset, team organization, and operation. Part 3 also offers guidelines for the content of team meetings and recommendations for meeting frequency, as well as how to run team meetings, when and how long they should be, how to make and maintain schedules, the content and distribution of minutes, the how and when to assign specific action items, and how to build consensus, team cohesiveness, common motivation and purpose, as well as pride in team membership.

Parts 4, 5 and 6 cover each of the three basic product development phases. Each part begins by defining the purpose, generic issues and team makeups for each phase. The generic steps to be addressed in each phase are discussed in detail, along with modern field-proven systems engineering methodologies to accomplish their timely and complete team based realization.

Part 4 is devoted to Phase A, Product Development Feasibility, and includes guidelines for determining needs, constraints, and resources for users, the developing institution, partners, and sponsors. The development of system concepts for pre-project feasibility planning is also covered. System concepts include development of a mission statement, design concepts, and operations concepts; identification of inherited equipments; risk assessment; top-level resource (schedule and cost) estimates; the selection of a proper system development paradigm; and a "first-cut" at a return on investment analysis (ROI). This part concludes with a discussion of the makeup and purpose of the Phase A review, which determines whether funding should take place for the next phase—Detailed Product Development Planning.

Part 5 covers Phase B, detailed Product Development Planning, which includes refinement and finalization of user needs, how to build and use work breakdown structures, and how to organize to staff the work breakdown structure. The importance of setting priorities to guide the concurrent development team so that everyone is working consistently toward common goals is stressed. Also covered is planning for establishment of control policies/plans for the management and evaluation of implementation, including development of review and reporting structures, technical margin management, documentation, risk control, configuration management, and supplier agreements. Construction of the resource plan, consisting of detailed, integrated schedules and costs based on the use of work precedence charts, is discussed. An outline for a comprehensive technical Product Development Team Management Plan is included. Finally, the more detailed ROI analysis required for Phase B is discussed, as well as the makeup and purpose of the Phase B review, which determines whether funding should take place for Phase C, actual Product Development.

Part 6 is devoted to the Concurrent Product Development phase, which begins with the staffing of the work breakdown structure, formation of the Product Development Team, and the development of formal signed-off requirements and design specifications. A sound methodology for requirements flowdown based on product priorities and mission product breakdown is presented. Also included is a detailed methodology for the definition and maintenance of system and subsystem interfaces. Modern traceable design trade-off methodologies are presented using product decomposition and priorities.

Also covered are the team's configuration management responsibilities, logistics support, and the impacts of inheritance, production, assembly, and testing requirements on product design, as well as lessons learned in testing. Basic progress review requirements are also presented. These reviews include the product requirements review, preliminary design review, critical design review, supplier reviews, test readiness review, pre-ship and post-ship reviews, and acceptance testing reviews.

That's it. Let's go.

Part 2

The What and Why of Teams

WHAT IS A TEAM?

There are more definitions of teams than there are teams. We speak here of teams used for three purposes: product development feasibility determination (Phase A), detailed product development planning (Phase B), and actual product development (Phase C). For these purposes, the following definition of a team offered by Katzenbach and Smith is by far the best one I have seen[1]:

A team is a small number of people with complementary skills who are committed to a common purpose, set of professional goals, and approach for which they hold themselves mutually accountable.

SOME WORDS ON THE WORDS

"A team is a small number of people. . . ."
That's right. Teams function best when they are comprised of six to ten good people. More than ten viewpoints on a single topic in any given meeting tends to bog down progress, becomes repetitive, and hampers efficiency. This does not mean that the "small" group cannot call upon outside expertise from time to time for support. A good team often does. It draws on outside help whenever needed to assure success.

". . . . with complementary skills. . . ."
Absolutely. Effective teams are comprised of people whose disciplines and experiences are related, needed, and, ideally, overlap to some

7

extent to foster communication and cooperation. You don't need or want redundancy on a team.

"*. . . . who are committed to a common purpose. . . .*"
There are at least two kinds of people. The first kind are worried more about how they are going to look to their bosses than they are about the success of a project at hand. Their loyalty vectors point upwards. The first kind you don't want. The second kind are people who put success of a product first and are more concerned about the team members than themselves. Their loyalty vectors point downwards, to where the people are who know how to get the work done. They are open minded, they listen, and they offer ideas without fear. If a mistake is made, they are glad to hear of it and accept correction because it advances the purposes of the team. This second kind of people are called *team* members, and they are found in organizations where management has set the tone. There is no fear. They are committed to a purpose. You want the second kind of people on *your* team.

"*. . . . who have a set of professional goals. . . .*"
Right again! The professional goal is to make the right decisions regarding go ahead, and if development is approved, to plan thoroughly and to execute those plans on time and within budget. The goal is not achieved by an individual. It's too complicated. The goal is achieved by teams made up of individuals. Nobody sinks a basket from half-court anymore. If the product wins, the team wins. When the team wins, the individuals on the team win.

"*. . . . and approach for which they hold themselves mutually accountable.*"
If you are a team member, or team leader, don't ever accept responsibility without authority. Good managers make clear what products they expect from a team and then they cut 'em loose. Good managers empower teams' personnel to be accountable. Accountability fosters self-respect and self-esteem. These are the number-one values coveted by everyone in every workplace. They are more important than salary.

CHANGING IDEAS

The concept of teams is not new. What is new is the growing discussion and formal writing on how to form teams, how not to form teams, and how teams in the industrial setting can be employed to actually improve our efficiencies and the quality of our end products. The current popular formalized discussion has been fostered primarily by the advent of the quality movement.

The culture, however, has not always nourished the team concept as we know it today. In fact, many managers are still in tune with the traditional industrial engineering concept of dividing work into small components and assigning individual workers, or organizational departments, to restricted specialized tasks.

This is not surprising. The "division of labor" concept has a significant and powerful historical basis. It was given formal credibility in industry through the publication of *Principles of Scientific Management*, by Frederick W. Taylor. Taylor espoused the reduction of each job into its most divisible parts, along with assignments of each part to individual workers—this in the name of efficiency. The "divide and assign" paradigm became the early model for mass production. It became the basis for time and motion studies, queueing analysis, and a whole bag full of industrial engineering lore. Not that improvements weren't realized; they were. But they were largely imposed by an elitist management that knew best what to do. With this mentality, workers and machines were basically viewed alike— as resources in the equations of improved efficiency. It was the way to do business. Taylor, Henry Ford, the United States, and everybody else loved it.

But some interesting things happened slowly throughout the second half of the twentieth century. Among them has been the emerging concept that workers may be more than mere biological robots. The lowest of the low, through experience, may actually be in a position to help us do better, to further enhance efficiencies, and improve the quality of our products. Significantly, the implementation of this idea has largely been through the medium of what we now commonly refer to as teams. From the early 1980s in the United States, the idea has steadily progressed, with multiple agonizing failures and sometimes spectacular successes. But why do some teams realize such triumphs and so many others bomb out?

SUCCESSFUL TEAM INGREDIENTS

Every good book on teams has at least one section on what makes teams work. This is mine.

Successful teams must have management commitment, experienced team leaders, proper training, well-defined products and processes, and individual and collective self-esteem.

". . . . management commitment. . . ."
This means top management, middle management, and line management. The commitment begins with the realization that a vast store of valuable knowledge lies within the personnel of an organization. It continues with a major paradigm shift away from bossing and toward leading.

Definition of a boss: An ego-driven individual whose uncertainty, lack of knowledge, poor training, and fear of superiors are exemplified by the issuing of unresearched, unilateral orders.

Bosses have an attitude. "I'm smarter than you are. If you are smarter that I am, then how come you're working for me?"

Definition of a leader: An informed individual who exhibits the peculiar capacity to communicate specific understandable goals, provide training and resources for their realization, and then delegate the authority to accomplish those goals through empowerment without fear.

Leaders have an attitude: "You are an important asset to this organization. Why? Because it is quite likely you know more about the day-to-day operation of your job than anybody else, including me. I'm listening. If you have better ideas, I will get us support to change."

Management commitment is not communicated through memos or company speeches or sitting in the office. It is conveyed through daily actions, by going to where the work is done, and asking questions. "What do you think?" "What are your ideas?" "How can I help you succeed?"

". . . . experienced team leaders. . . ."
Good team leaders have ample experience across all disciplines of the team's charge, enough to recognize what they don't know. We are most dangerous when we don't know what we don't know. If we are aware of what we don't know, we can get help from people who do know. The team leader doesn't have to be more knowledgeable than everyone on the team. If he or she were, you wouldn't need a team. But sufficient knowledge is required to anticipate the pitfalls and ask the right questions and to earn over time the respect of all members. Good team leaders also know how to listen and to take opportunities to build the self-esteem of all individual team members. They have technical skills and are skilled in human relations as well.

". . . . proper training. . . ."
Recall, we speak here of teams used for three purposes: product development feasibility determination, detailed product development planning, and actual product development.

The people within your organization selected for these teams should already be well trained in their disciplines. We are not dealing here with team training in such specific areas as statistical process control, or continuous improvement in production, or more general process identification and improvement teams that involve brainstorming.

In the context of our topic, team members should receive training on their individual accountability for the processes and products to be covered in this book, and they should have communicated to them the tone of confidence and authority invested in them by upper management. This can often be achieved during initial team meetings by the team leader. The first meeting, however, should always include representation of top management to establish the team's charter and the team leader's authority. I'm going to say this more than once: "Don't accept responsibility without authority!"

Selected team leaders often benefit from training in team leadership dynamics. Sources for team leadership training are books, consultants, and a host of available seminars.

". . . . *well defined products and processes.* . . ."
This book is about products and derived processes to successfully develop those products. Note, we do not start with inventing processes in a vacuum. We start by determining the products we require and then developing a process to realize each product. Product determines process. If you have a process in place that has no accountable product, chances are that it is out of control or it is not needed.

Parts 4, 5, and 6 of this book present checklists for the products and processes for each of the exercises of determining product development feasibility (Phase A), detailed product development planning (Phase B), and actual product development (Phase C). Each of these involves both intermediate and end products. The end products of the feasibility stage and the detailed planning stage are reviews given to upper management that include findings and specific recommendations. The end product of the actual development stage is a product that is meant to be bought or accepted by a final user or set of users— generally the consumer.

An intermediate product is one that is created as a result of a well-defined process carried out in direct support of an end product. Intermediate products consist of documents or physical entities. Thus, items such as investment analyses, mission statements, risk assessments, schedules, requirements, specifications, reports, test rigs, prototypes, subsystems, meeting minutes, and interim reviews are examples of intermediate products. The successful team leader must provide a clear statement of the products for which the team is responsible and the processes to be followed to bring them to fruition. Remember that people thrive and perform best when there is good structure. The checklists in this book are intended to assist in creating that complete structure.

"*. . . . and individual and collective self-esteem.*"
Does this sound a bit endearing? Let me tell you, if you have to pick one attribute of a successful team that is the most important, no matter what the team is doing, this is it.

Not a lot has been written on the application of teams to the activities covered in this book. But a whole bunch has been documented on what happens when human beings are provided with an atmosphere in which they can feel challenge, pride in what they do, a sense of professionalism, and a feeling of importance through meaningful contribution. It's called self-respect. Is this true? In 1991, the American Productivity and Quality Center found that challenging work (empowerment) was ranked as the most important job-motivating factor. Recognition (pride) for a job well done came in second. Pay? It ranked fourth.

Tomes can be written on this alone, and I won't belabor the point. Allow me, however, to provide a few random examples. They go back a bit; that's what's good about them.

Welcome to the Mines

In 1963, Eric Trist and his research colleagues referred to the Taylor management approach as "conventional"[2]. Their work included studies of the application of the conventional management approach to the British mining industry. Work was divided into detailed functions that were assigned to individuals. Quality control was one of these separate functions. The conventional paradigm, in theory, is a convenience for management in which the supervision of workers is reduced to the simple monitoring of intelligent animals engaged in simple repetitive tasks.

What makes the Trist work so important is that the researchers also studied a second organizational structure employed in neighboring British mines—the so-called "composite" structure. In the composite arrangement, teams of miners were formed and made responsible for the total task of coal extraction. Significantly, membership in the teams was determined by the miners themselves, much as a group might choose up teams for a sandlot ball game. The teams managed themselves in a given workplace, including the distribution of wages among themselves. The team was also responsible for its own quality control.

Trist reports remarkable differences in productivity, attitude, and quality between the two concepts of worker organization. The composite work group was highly organized and stable. It was no longer unrealistic for individuals to attain stations of increased importance within the group. The concept of being asked to perform more than a simple task was no longer regarded as exploitation. The opportunity to learn additional skills was actively sought. Supervisors of the groups migrated from issuing orders to simply providing technical advice. They became "leaders" instead of "bosses." They became good guys.

Teams identified problems and were motivated to analyze them and work out ways to fix them. Change was accepted willingly. Attitudes in the workplace transformed dramatically. Absenteeism declined by a whopping 60 percent. An overall improvement of 20 percent was reported, as was an incredible 42 percent increase in productivity. The groups had social meaning and clear goals. Individuals had self-esteem: "I am needed here."

Will It Play in Ahmadabad?

The Trist research group next sought to test the validity of their discoveries in the mines. A similar study was conducted by implementing composite work groups at the Ahmadabad textile mills in Ahmadabad, India. The culture was entirely different, and the workforce made up of females. The results were quite similar to those previously observed. Improvements in quality of 30 percent were especially significant.

Non-Linear Systems

In 1963, similar results were reported as a result of the implementation of composite work groups at Non-Linear Systems, Inc., a United States West Coast manufacturer of electronic test equipment. Again, control of organization, production, test, and quality was taken over by teams. Over a two-year period, productivity was reported to increase by 30 percent, with a decrease in customer complaints of 7 percent. Improvement continued constantly over time as the groups were persistently self-motivated to identify problems and devise better methods of operation. Quality continually improved to the near elimination of defects. The role of inspection in quality management greatly diminished.

FMC Truck Assembly

The Ford Expedition people understand what full concurrency means. Close to 100 workers on the assembly plant floor gave inputs to the

Expedition design. Examples of improvements included in-line vehicle sequencing and the "just-in-time" arrival of instrument panels. The Expedition team understood that the real experts are the workers and that leadership involves organizing that knowledge through listening and building motivation. They understood that the greatest compliment one human being can offer another is to simply ask, "Can you help me with this? What do you think we should do?" When serial number 1 came off the floor, everyone went outside and, amid much backslapping, took pictures of themselves alongside the vehicle. Then they went back to work—all smiling.

Fiero in the 80s

As a result of the 1980 airing of the NBC documentary, "If Japan Can, Why Can't We?" Dr. W. Edwards Deming was invited to Pontiac. The resulting impacts on the development of the Fiero were substantial. Close teamwork—conventionally alien to Detroit—between design, engineering, manufacturing, assembly, sales, marketing and all aspects of development was established. Significantly, suppliers were involved in designs—some before contracts were even finalized. Worker classifications were reduced and responsibilities delegated. Worker allegiance and devotion to the new car flourished. Said one Fiero manager[3]:

The management has to take the first step to create the proper environment. At GM we have a culture that says we're in this to make money. But we have focused too narrowly on the end result, instead of recognizing that we're dealing with people with very broad desires. We've become managers, not leaders.

Fiero won awards—Best 1983 Design, American Car of the Year in 1984, and another rating among the Ten Best of 1984.

Cadillac in the 80s

In the late 1980s, Cadillac was on the move. Cadillac's Roseta Riley recounts the story[4]:

We downsized in the gas crisis despite what customer's said. We dropped to 16th. We were entrenched in traditional behavior with adversarial relations with our people. We instituted cultural change, a new focus on the customer and a disciplined approach to planning. We had no quality plan. The business plan is the quality plan. From 1985 to 1990 we came from 16th to 4th.

In 1985, Cadillac implemented "Simultaneous Engineering" involving cross-functional teams from engineering, manufacturing, suppliers, finance, and the United Auto Workers. A two-year basic training program was instituted.

One team, the Recognition and Rewards Team, decided to eliminate rankings and employee appraisals, explaining that "One half is above average and the other half is below average. Tell people that they are average long enough and they will be. We don't have 'average people' anymore—we have quality people."

In 1987, Cadillac formalized a "UAW/GM Quality Network" team under which joint responsibility for quality was accepted by the union and by GM. Traditional adversaries became partners. One UAW part design saved a whopping $52 million.

Cadillac turned the organization chart upside down. Managers stopped bossing and became leaders. As Riley puts it, "Management set strategic objectives and people decided how to get there."

"Design For Manufacturability" teams led to simpler design, ease of assembly, and reduction in variation. Among the results were a 70 percent reduction in customer problems, a 65 percent improvement in reliability and durability, and a one-year reduction in lead time.

When the Malcolm Baldrige Award people asked a 26-year veteran Cadillac worker how he liked his new team-oriented job, the answer was, "The first twenty years were awful. The last six have been fantastic."

Cadillac turned around because the bosses turned around. The turnaround set an atmosphere for the latent power of respect and teamwork to flourish.

The multidisciplinary team concept exemplified at Fiero and Cadillac is an integral component of the quality movement. The expanded team concept is catching on. It is now finding its way into conventional systems engineering practices in the form of additional key membership on what systems engineers have traditionally called the system design team. In the past, system design teams have customarily focused on hardware and software design throughout product development. An increasing number of development product design teams are now including marketing, production, suppliers, operations personnel, and customer representatives, in addition to simply hardware and software design personnel. When total teamwork flourishes among people of all departments, it is not surprising that traditional adversaries became compatriots.

Hewlett-Packard

Tom Peters likes to ask questions. He once asked a woman on an elevator where she worked. She replied, "Fourteen." That's where she got off.

Contrast that response with one from an electronic parts packer in a shipping department at Hewlett-Packard. An ex-employee of HP, and now a colleague of mine, related to me what happened one day when she asked a parts packer where he worked and what he did. What followed was a 15-minute unilateral discourse explaining the computer-supported fault isolation techniques used by field engineers and how that translated into requests for specific parts to be shipped to specific locations in a timely manner. The packer related in detail how it was his responsibility to support the customer by getting the right parts to the right place at the right time without error. He walked her through the paces of receiving an order, getting the part, packaging it, and placing it in the right-sized box with appropriate protective filling. Next, he made up the label and placed it on the box. Then, he looked inside the box again to be sure the content was consistent with the order and the label. Next, he held the box closed and shook it to be sure everything was safe. Then, and only then, did he close and seal the box.

I do this very carefully but without delay. If the right part doesn't get to the right place at the right time, the customer ain't happy. If the customer ain't happy this organization is in trouble. It's very important that I don't make a mistake. I know there's a lot of people in this company, but if it doesn't happen right here—the whole thing isn't going to work. I'm an important part of this whole team. A lot depends on what I do right here. That's what I do.

Why does the HP worker give an answer that lasts a full 15 minutes, whereas the first response of our friend on the elevator to the question of where she worked was location, not function? Simple. He has a team mentality. The woman who got off on the fourteenth floor works *for* a *boss*. The parts packer works *with* a *leader*.

The Mars Rover in the 90s

Most of us are aware that money for big science is getting tighter all the time. At NASA, the faster, better, cheaper paradigm is in place. The billion dollar flagship spacecraft of the Voyager and Galileo class are beginning to be replaced with outer planet exploration budgets on the order of 250 million. Enter the Mars Explorer, which has recently excited us all. Donna Shirley, at the Jet Propulsion Laboratory, put together the original team. The new concurrent team was given complete responsibility to meet meaningful science objectives at one quarter the cost everyone had been used to. Any and all crazy ideas were brainstormed. It was nutty and immense fun. Among the problems was that of landing. There was not enough money for a rocket-controlled backdown, nor for a lone parachute drop that would involve high g forces at impact. The answer was a combination of a chute and a shock-absorbing container that in fact bounced along the Martian surface almost 10 kilometers before coming to rest. It worked. Cost cuts were everywhere. Commercial-grade equipment was used anywhere it was possible. The entire ground operations team fit into a single room for the first time. Innovation and excitement abounded. The system remained operational long after its 10-day mission design—to the delight of everyone. There was no question that the rover and its mother spacecraft were designed and produced by a fully multidisciplinary team with a high degree of individual motivation and leadership. Not a single member of that team will ever forget as long as they live their collective accomplishment. It was their baby. Talk about pride of ownership and self-esteem! They had it good.

Socrates Then and Now

Unfortunately, Socrates didn't write very much. But Plato wrote of him extensively in the *Republic,* in which he speaks to us of thymos. In our own century, Fukuyama picks it up nicely[5]:

Thymos is something like an innate human sense of justice: people believe that they have a certain worth, and when other people act as though they are worth less—when they do not recognize their worth at its correct value—they become angry (anger is the literal translation of thymos). The intimate relationship between self-evaluation and anger can be seen in the English word synonymous with anger, 'indignation.' 'Dignity' refers to a person's sense of self-worth; 'in-dignation' arises when something happens to offend that sense of worth. Conversely, when other people see that we are not living up to our own sense of self-esteem, we feel shame; and when we are evaluated justly (i.e., in proportion to our true worth), we feel pride.

■■■■

A simple message: Most good ideas are simple. Yet, it contains some of the most important words in our lives. Fukuyama doesn't take them lightly. He contends they are the expressions that drive all of humanity—and consequently *all of history itself.*

Notice the absence of the words "power" and "money."

In today's culture, power and money may be mistaken as a means to self-esteem, but they are not what it is all about. Power and money are not necessary for self-esteem. Fukuyama's words need to be remembered when we think about workers: justice, worth less, worth, angry, indignation, dignity, self-worth, self-esteem, shame and pride.

So what's new? Evidently, what makes teams work goes way back.

Now I am not about to suggest that tools such as SPC, QFD, and XYZ are not valuable and of significant importance. They are. But I am suggesting that when a host of individuals have self-esteem through an opportunity to contribute, when they feel important, when they care—that's when quality happens. People, not tools, do things.

Forward Again

The rest of this book will discuss the application of teams in a slightly different way from the examples above. In the broad sense of the word, teams are already routinely widely employed for product development feasibility determination, detailed product development planning, and actual product development.

What we will explore are ideas based on hard-won experience related to the cross-organizational makeup of such teams, the completeness and correctness of their products, and the processes and methodologies they employ. I repeat, the intent is not to tell you how to do business. It is rather to provide ideas on effective concurrent team use oriented toward attaining completeness of content and structure—in short, to assist in the constant improvement of the product development process.

ENDNOTES

1. J. Katzenbach and D. Smith, *Wisdom of Teams—Creating the High Performance Organization* (Boston: Harvard Business School Press, 1993).

2. E. Trist, G. Higgen, H. Murray, and A. Pollock, *Organizational Choice* (London: Tavistok, 1963).

3. Andrea Gabor and Jack A. Seamonds, "GM's Bootstrap Battle: The Factory Floor View," *U.S. News and World Report,* 21 September 1987, 52.

4. R. M. Riley, "Cadillac Motor Car's Turnaround." Luncheon address presented at the Second International Symposium, International Council on Systems Engineering, Seattle, Washington, July 1992.

5. F. Fukuyama, *The End of History and the Last Man* (New York: The Free Press/Macmillan, Inc., 1992.)

3

Team Leadership Guidelines

We will now turn to the importance of management support, the role of systems engineering in team operation, and the requisite mind-set of the team leader. This section also provides a few hard-won ideas and guidelines for team formation and operation that apply to all three product development teams: the initial product development feasibility team, the detailed product development planning team, and the actual product development team.

WHERE IS MANAGEMENT?

Do not attempt to take on team leadership responsibility without the complete and thorough backing of management. I've said it before, and I'll say it again. It is that important. This means top managers overseeing product development, managers of the people you want on your team, and your own immediate manager. Everyone must understand what you intend to undertake. And they must delegate not just the responsibility, but the necessary *authority* for you to succeed. Authority means you have the power to make significant product design decisions and recommendations. This should not be threatening because adequate management reviews will take place before anything is finalized. If this is not clearly understood, give them a copy of this book. If they still don't get it, look for something else to do. But it's not likely to be that bad. Most managers today are pretty good, and getting better. The point is, they must be in your court, and everyone has to know they are in your court.

THE ROLE OF SYSTEMS ENGINEERING

The discipline of systems engineering is the foundation for team based product development. The term "systems engineering" is used throughout this book, and a common view of what it means in the context of product development is essential.

Systems engineering is defined as:

The systematic application of proven standards, procedures, and tools to the organization, execution, and control of product feasibility studies, detailed product development planning, and the development of product requirements, design, fabrication, integration, test, and logistics support.

Some words on the words.

Systematic Application

Systems engineering involves a definite generic sequence of activities that must be faithfully adhered to. These are covered in detail in Parts 4, 5, and 6. In general, the earlier stages of each sequence are the most critical. Serious attention should be given to potential errors of omission.

Proven Standards

There are many valuable standards that have been widely used and revised through practical experience. There are many useful sources such as professional societies, institutes, and the military. (Individual organizations often produce their own standards.) Good standards are of significant worth, even when not formally imposed on a specific project, because they can serve as a guide in assuring that all aspects of an issue have been covered. Almost every question you may have regarding standards has been faced by somebody before you. The library of existing standards can serve as a valuable source for ideas, guidelines, and solutions.

Procedures

These refer to the actual steps to be carried out in each of the product development phases covered in detail in Parts 4, 5, and 6. Procedures include development of user needs, product concepts, control policies and plans, requirements, specifications, and all the rest.

Tools

This is the set of intellectual and physical tools used to support product development. Intellectual tools are methodologies for trade-off analysis, configuration management, requirements flowdown, and so forth. Physical tools are computer-aided design, electronic documentation systems, mockups, prototypes, and the like.

Organization
This refers to the hierarchical and complete staffing of the work breakdown structure items. Generic top-level items consist of management, systems engineering, product cognizant engineering, and test and logistics support.

Control
The act of systematic application of proven standards, procedures, and tools to product development.

Product Feasibility Studies
These studies address questions to be answered in determining top-level practicality of product realization and its market viability. That is, is it a good idea? These studies are conducted during Phase A (see Part 4).

Detailed Product Development Planning
If the product idea proves feasible, approval to carry out detailed product development planning is granted. Detailed planning consists of complete preparation for all activities to be carried out in support of the actual execution of product development. This planning is conducted during Phase B (see Part 5).

Development of Product Requirements, Design, Fabrication, Integration, Test, and Logistics
These steps support actual product development and are carried out upon approval of the detailed product development plan. These activities are finalized and carried out during Phase C (see Part 6).

TEAM LEADER MIND-SET
The familiar term "interdisciplinary" is used to refer to the mixing of totally different fields of endeavor. Today the term is even more complex. For example, the single field of engineering is so diverse that it has even become interdisciplinary within subfields. Consider the vast differences between computer science professionals specializing in software development, VLSI fabrication, bus design, quality control,

networking, training, and so forth. It is not reasonable to expect that a single person know "everything." It is, in fact, impossible for a single person to know everything. Nonetheless it is true that team leaders must *know what they don't know.* When we don't know what we don't know, we are dangerous!

The lesson is that we must try very hard to understand the boundaries of our knowledge. Team leaders—engineers in particular—must take the time to learn enough about every technical discipline that affects them in order to recognize when to ask the right questions or to get help. There is nothing wrong with limited knowledge. The team leader is not expected to know everything, but the team leader is most certainly expected to be aware of everything.

It is important for a team leader and all team members to understand that the real in-depth knowledge lies in the areas of expertise of each of the team members. The soccer ball analogy is helpful to bear in mind. As presented in Figure 3.1, visualize the product, or the system, as a soccer ball. The external circle of the ball itself is the product boundary—the boundary that the product user sees and interacts with. The internal seams on the ball are the subsystem boundaries, that is, the subsystem interfaces. Note that some of the subsystems have external interfaces as well. A team leader is basically doing what systems engineers traditionally do. They are responsible for walking the outside and inside lines (seams) of the soccer ball. Subsystem cognizant engineers (COGEs) are responsible for meeting requirements on their interface boundaries. All members must understand some-

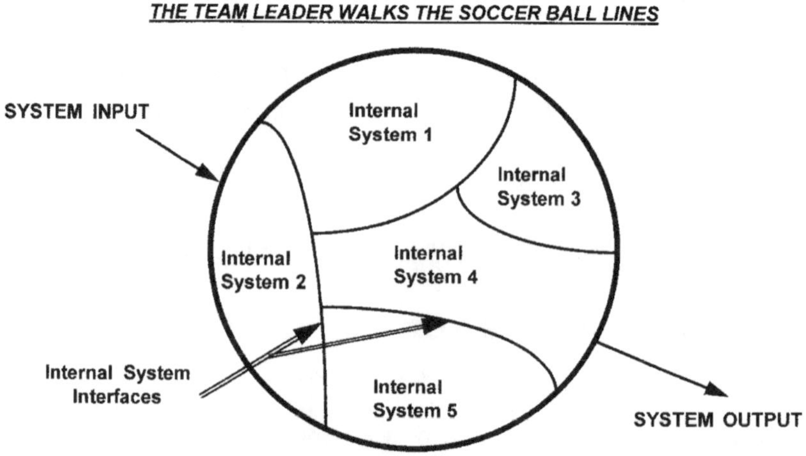

THE TEAM LEADER WALKS THE SOCCER BALL LINES

Figure 3.1. The soccer ball analogy.

thing about how they intend to do this because it may effect other subsystems. The point is that it is basically up to the subsystem cognizant engineer to decide the details of how his or her subsystem is going to work. The team leader must strike a balance between knowing what's going on and micro-managing. That is, as much as possible, stay out of the COGE's face.

Since the team leader is responsible for the performance of the product boundary, it is important to keep in mind that optimization of each subsystem performance does not necessarily lead to optimum system performance. The team leader should make this clear to the whole team. Subsystem requirements are derived from total product requirements—the outer rim of the soccer ball. These requirements, no more and no less, are the ones to be met.

In this and all other team related issues, a primary goal of the team leader is to gain a consensus among team members, a consensus that is consistent with the current and long-term goals of the team. Consensus is required with regard to all team activities including, WBS construction, schedule making, interface definition, selection of design criteria and priorities, development and documentation of requirements and specifications, identification and execution of special studies and options analyses, action item assignments, choice of design tools and aids, assignment of failure reports, assessment of user feedback, lien corrections, and all other pertinent business of the team that the team leader deems mandatory.

Consensus is best obtained by clearly asking for it. For example, when discussion reaches a point at which a conclusion seems imminent, it is wise to reiterate the conclusion in summary by stating, "We all agree, then, that . . ." and to probe those members of the team whose disciplines are most affected. If discussion is not leading toward a consensus, the team leader should immediately try to determine which of two possible conditions exist: Consensus can be reached in the framework of that particular meeting, that is, within five or ten minutes, or it can't. If it is felt that consensus is possible in a reasonable amount of time, it can usually be extracted by guiding the discussion through asking questions. The questions should be specific and directed toward the consequences of unacceptable proposals or actions under discussion.

If it is felt that consensus on an issue cannot be reached in the time frame of the meeting, either because of uncertainty as to what should actually be done or because of lengthy discussion, then an action item should be assigned to an appropriate individual to study the issue and recommend a course of action at the next meeting. This not only gives the team leader (and everybody else) an opportunity to think upon the matter further, but it also provides an opportunity to discuss the issue in private with the assignee over the period of the next week(s). This approach stresses that team meetings are not working meetings. They are meetings whose purpose is to review status and to identify issues. The work is done outside the team meetings.

Ideally, team meetings should not last much more than one hour, although they may on occasion last up to two hours. Within this time frame, if consensus on a given issue is not reached within ten to fifteen minutes after sufficient discussion has taken place, it is likely that it will not be reached during that particular meeting.

Again, team meetings are not working meetings. They are rather geared toward a review of status, the identification of system-level product issues, the assignment of action items, and the maintenance of team coordination and communication. Issues that are not resolved in a timely manner are best delegated and addressed outside the design team meeting in routine working meetings that focus on those single issues.

The team leader is particularly challenged to exhibit leadership. Once again, there is a difference between being a boss and being a leader: Bosses give orders; Leaders do just that—lead. They lead by identifying and drawing out the best qualities in people and focusing those assets on problem identification and solution. Bosses generate fear. Leaders do not generate fear; they generate cooperation. Leadership entails the ability to listen, to involve people who know more detail than you do and to allow them to achieve pride through their contributions and workmanship.

The good leader earns respect over time. Respect is earned through technical thoroughness, giving of respect to those more knowledgeable, understanding and sticking to the generic implementation process, and building human beings by focusing their best abilities toward a common, well-defined purpose.

Here's a team leader mind-set summary:

1. In most companies, the loyalty vector points upwards. This is called fear. In quality companies, the loyalty vector points downwards. That's where the knowledge is to get the job done.
2. Invest the time to sharpen your skills. You must earn leadership through respect.
3. Worry about the customer more than you worry about yourself.
4. Support each team member's contribution. Build a team atmosphere.

5. Give whenever you can. Separate the important from the unimportant. Don't quibble about the unimportant. Amass perks.

6. Lead, don't boss.

7. Don't pretend to know it all; you don't.

8. Avoid being negative. You don't have to say "No," or "We can't do that." Express yourself positively. Let the team function. Build people.

9. Be Socratic. Ask questions to gain understanding and consensus.

RUNNING THE MEETINGS

The following paragraphs cover guidelines for team meeting frequency, preparation, co-location of team offices, meeting content, and team meeting minutes, with a brief comment on buzz words.

Meeting Frequency and Timing

Team meetings should take place once each week. Teams that do not meet once a week or that meet "when issues arise" are not the teams we are talking about. They are some other kind of team.

The selection of the particular day can have advantages. Fridays have the disadvantage that people tend to lose a certain steam over the weekend. People also generally go through a transition start-up period on Mondays and gain momentum as the week progresses. If they have received an action item assignment during a Friday meeting, they are more likely to lose some of the crisp detail of what must be done between Friday and Monday. On the other hand, conducting meetings on Wednesdays enhances the probability that work on actions will begin during the latter part of the week, when productivity is still on the rise.

Avoiding Monday also provides the team leader time to prepare a productive setting for the meeting. It is useful to informally move among the team members on Mondays and Tuesdays and make a point of discussing issues that may be controversial, or even routine, sometimes one-on-one with specific individuals and other times in small groups. If it is sensed that two people have opposing views or approaches, the issues should be discussed individually with each to understand their viewpoints in an effort to define a middle road that does not violate the goals of the team leader. A basic goal in this approach is to find solutions prior to the team meeting.

This is not to suggest that issues should not freely be raised and discussed during team meetings. The goal, rather, is to minimize outright conflict and create a general meeting atmosphere of civility and cooperation. This is done by identifying and resolving serious conflicts outside of the meeting. Design issues can be the source of heated discussions, some technically oriented and many simply personality oriented. It is useful to adopt the approach of trying to anticipate delicate design and personality issues in advance and tackling those issues outside of the meeting forum. There is an important psychology to this. Civility breeds civility. It sets the tone for the team leader's command. Open refusal to adopt a proposed course of action, especially during team meetings, can rapidly diminish one's credibility as a leader. It is quite feasible to reduce the probability of spontaneous rejection at team meetings merely by minimizing surprise. The Monday and Tuesday preparatory minimeetings, or discussions, provide a significant opportunity for the team leader to obtain support for the direction and assignments he or she intends to put forth in the actual team meeting.

In these premeeting discussions, it is not wise initially to dictate. A more productive approach is to make the immediate goals clear and then ask, "How do you think we should go about this?" or "What do you think of this approach?"

It is also useful to discuss the progress of assigned action items. This affords the opportunity to discuss in advance any implications for other team members who may be affected. It is expedient to make such "rounds" on Mondays and Tuesdays as needed, which allows for the preparation of the terrain for a well-organized meeting in which direction occurs smoothly because the principal parties are informed. People respond very well to this treatment, as it is always a compliment to be consulted in private, or in a small group in advance.

Preparation of this kind typically results in nods of agreement as opposed to surprise or shock at the weekly meeting. Many people are tentative in embracing new ideas on the spur of the moment. Surprise and the often lengthy discussion that follows are generally to be avoided. Establishing an atmosphere of smooth cooperation is very important early in the game. If initiated at the beginning, the cooperative tone rapidly becomes a part of the group's behavior. Normally, antagonistic people (the ones who shoot from the hip) become highly civilized in such a cultivated cooperative setting. Major public disruptions during team meetings can drastically erode the team leader's credibility. If they continue over any period of time, the leader's effectiveness will rapidly degrade. The intent here is not to mesmerize creative people. On the contrary, the aim is to provide time for thoughtful consideration of identified issues. In this scheme, much of the effort by the team leader to direct the overall process often occurs outside of the weekly meetings. Being a team leader is a full-time job.

Co-Location

This one is very important. By "co-location" of a team I mean that all permanent members of the team have office space in one physical place with a common meeting room, secretaries, coffee makers, copy machines, and everything else. There are good reasons for this.

One is that people on the job tend to socialize with those in their immediate proximity. They talk about what happened last weekend. They get to know something about each other beyond the job setting. They have a chance to smile at each other and say, "Good morning." Friends are made. This is good.

Another important reason is that when a thought occurs to someone and he or she wants to talk about it with Eddie, it is easy to just walk up or down or across the hall, and there's Eddie. One doesn't have to make a phone call, get voice mail, and wait hours or more—maybe until tomorrow—to talk about it on the phone. It's much better face-to-face anyway. You can draw pictures on a blackboard, look at documents, and communicate far better, easier, and much faster. This is very important over time.

Another great reason: The team leader has a much better feeling about what everyone is doing every day. He or she is right there, totally accessible; things happen faster.

And, of course, the secretaries (or their administrative equivalent) have a total picture of what is going on because it all happens in one place. They too are coordinated with the team view. They are working with the team. Their efficiency and performance are necessarily improved through co-location with the team.

The idea of co-location is not as inconsequential as it might sound. It has definite day-to-day advantages. (I seriously recommend it.) Ask Chrysler what co-location of design alone has done for their time to market. I often say there are only three things to remember regarding team location. Co-locate, co-locate, and co-locate.

Team Meeting Content

The very first meeting of any team should be devoted to

1. a review of the team's mission
2. a review of the process to reach those goals

3. a review of expected products and a schedule for their accomplishment

4. an explanation of why each team member is present and how their contributions will map onto specific products, for example, "Charlie, you are here because of your knowledge of. . . . We are looking to you to. . . ."

5. a review of team meeting frequency, location, and content

6. a pep talk on how we have been empowered by management; our reliance on, recognition, and respect for each other; our individual importance; and our professional responsibility to work together toward success on time and within budget

This is a critical opportunity for the team leader to confirm and establish his or her technical leadership ability. When teams are first formed, there is a transient period of internal adjustment. Everyone is feeling everyone else out, and there is a tendency for individuals to seek to establish, or authenticate, their presence before the new group. The team leader must be sensitive to the fact that there is always a period of initial personality adjustment during which people may speak for the purpose of speaking, independent of specific subject matter or group goals. Basketball teams succeed by moving the ball down the court as close to the basket as possible to enhance the probability of improving the score. That takes group coordination. They do not succeed by having individuals constantly trying to sink one from half court. An early tolerance is often required—mixed with a firm, but not overbearing, preservation of goals.

Each subsequent meeting should begin with a review of lower level schedules of the operating plan. Expect revisions to be called for. If weeks go by and everything remains "on schedule," it probably means the schedules are not detailed enough and there is a risk of loss of visibility. It is also possible that the team leader is not being told everything that is happening.

Alternatively, if line items are constantly changing, it is likely that the schedules are too detailed. In discussing schedules, address the time frame of one to four weeks around the present date in sufficient detail. If a team member asks for another week or two to complete a specific line item, be ready to give it. But also discuss the impacts on other line items, and formulate approaches that will maintain the schedule for the item due date if needed. Changes in line item end dates on a schedule may easily affect other items on that same schedule or the schedule at the next level up. The working schedules under discussion at the team meetings should be below the level of schedules routinely reported to management. Thus, the team leader has his or her own built-in cushion and flexibility on week-by-week schedules. Of course, the importance of meeting end dates expected by management must be stressed as a team commitment.

A few more comments for the team leader. Make sure the work scheduled for the following week and month is the work that is actually being done. If it is not, line items should be added as required and other line items adjusted appropriately. It is important to strive to understand the reality of what is happening, but try to understand this reality without issuing reprimands. Everyone should be comfortable with the schedule. Stress that you are asking for professional commitments. Spend the time to review every item. Ask your questions and then be quiet.

Some years ago, my line manager came to me after sitting in on one of my product development team meetings. "Do you realize," he asked, "that your meeting lasted one and a half hours and you were talking a full hour of that time?"

"Was I?" I said, slightly amazed.

"They all know what you're thinking," he said, "but do you know what's on their minds?" He helped me realize that you don't learn by talking.

Draw the team members out as much as possible. Listen. If you sense that a serious slippage is possible, beyond the limits of the team's time cushion, ask their advice on how to recover. Make every effort to implement their plan for recovery. Move other line items and make temporary shifts in personnel work assignments. Involve the whole team in attacking the problem. Stress that the discovery of a problem is a positive event. A constant goal in scheduling is to discover problems as early as possible—weeks or even months in advance. That is, in fact, why schedules are made.

Discussion of the status of action items should conclude with a reiteration of which individual is responsible for its closure and the validity of the target date for closure. A single individual should be responsible for closure of each action item, even though that individual may lead a subgroup to perform the actual work. Assignment of an action item to an individual need not always mean that a specific person will be doing any, or all, of the work. He or she may complete all of the work, do part of the work with help from others, or totally delegate the work. The important point is that more than one person cannot be responsible for action item closure. When appropriate, the team leader should also take responsibility for action items.

In the final phase of the meeting, it is useful to poll the team members individually, left to right. This provides a well-defined opportunity for team members to raise additional concerns and issues. Simply asking the group, "Is there anything else?" is not sufficient to draw out information in the detail required.

For example, your conversation might go like this:

"Jane, do you have anything else?" Then be quiet and wait. If Jane says, "No," then say something related to her efforts. Anything.

"Do you still feel good about the schedule?"

"Yes."

"This line item six doesn't bother you?"

"Not really, if anything it's number five."

Bingo. There *is* something on Jane's mind. This is a good result.

The team leader doesn't have to do this with every team member at every meeting. Some people need very little prodding to talk incessantly. The others who tend to remain quiet over long periods of time are the ones you need to encourage to participate. Often they know more about what needs to be done than anyone else, including the team leader! Use your sensitivity and judgment. Much of this has to do with the individuals you are communicating with, what they are currently doing, and your own confidence in how the work is progressing. But definitely address each person around the table at each meeting, and pause noticeably, or draw him or her out, to provide a chance to respond. If you don't communicate, you're dead.

Once in a while, somebody comes up with a dumb idea. You may even do that yourself. I sure as shooting have. If this happens, and you recognize it, you don't have to be negative right off the bat. Ask other team members at the table who may be affected, "Interesting, what do you think of that, Marty?" It is informative how this Socratic approach gives rise to finding the fault with an idea. Let the team be a team. Remember, they are smart people. When possible, it is better to have the team come to conclusions than to dictate, even though in the end you have 51 percent of the vote.

One more comment on this left-to-right business. One of the more interesting Phase C product development teams I have been privileged to lead involved a sophisticated implantable medical device. We had a bright assemblage of MDs and PhDs—all specialists in materials rejection, communications, control theory, VSLI, and so on. After our first few meetings the format was well established and everything was smooth and productive. Around our sixth week, a new hire specializing in sensors came aboard. At his very first meeting, I started out by introducing him to those he had not met. The new member then went off on a discourse of his thoughts on the role of sensors for our product. I noticed that everyone was a little uncomfortable, and I sensed from glances that came my way that the rest of the team would not mind at all if I interrupted our new member and proceeded with our regular format. But I didn't. When he finished, I said, "Those are interesting ideas," and began our meeting with schedules as usual.

At the very next meeting, our new member understood the format and knew he would have his chance when I went around the room later in the meeting, as usual. The point here is that you can often accomplish what you want by being a bit patient. In this case, I purposely avoided chastising our new and enthusiastic member at the outset in front of the team. All I had to do was adhere to the structure and wait a bit. The message is that the team leader sets the tone not only for format, but for civility, respect, and everything else by example. Doing is a very effective way of communicating.

Your team is your lifeblood. To the extent that you lose touch with team members, you lose touch with reality. Your ability to communicate is imperative.

Team Meeting Minutes

The team leader may designate a team secretary to generate minutes or may generate them personally. The generation of minutes by the team leader has specific advantages. An example of design team minutes to be issued after every weekly meeting is given in Table 3.1.

First, it provides an opportunity to mention individuals' names as often as possible and include everyone's name over a period of time. The team members' supervisors generally read the minutes, and it is a subtle way to give everyone points.

Second, it is a practical way to control the honesty of what is reported. When good things happen, they should be conveyed to management. When a problem arises, it should also be noted, along with the plan for action. If a problem persists, management then has some visibility. Surprises are almost always undesirable for everybody.

Third, the minutes serve as a valuable audit trail for team progress and the rationale for that evolution. Minutes are a product.

The text portion of minutes should rarely run in excess of a single page. Each item covered represents a point worth recording in the context of the overall flow of development. The issues are technical, and administrative issues are seldom included (these are covered in staff and project meetings). Results of action items and special studies

---

▐⫶ TABLE 3.1 *Sample System Design Team Minutes*

Date

TO: Attendees and Distribution

FROM: (Team leader's name)

SUBJECT: Minutes of Project X Design Team Meeting of (date)

ATTENDEES: J. Brown, N. Green, J. Grey, M. Maroon, R. Opal,
                   R. Purple, G. Red, (your name)

NEXT MEETING: Wednesday, (date), 10:00 AM, Bldg. 204, Room 128

Development schedules were reviewed. All items remain on schedule with the following exceptions: (1) Procurement for the X-2000 terminal has been delayed for an estimated two weeks due to a vendor backlog. (2) The scheduling of coding for software module FOO will be interchanged with software module BOO, which is not dependent on the X-2000. (3) A line item for software conversion to accommodate system queueing for the new plotter will be added to the level six schedule. There are no schedule slippages anticipated at level four as a result of these changes.

G. Red reported on results of disc access simulation studies. A recommendation to upgrade the operational system disc unit to model 502 was made as a result of these studies. The recommendation was accepted by the team. Action item No. 23 is closed.

It was agreed to extend the current development software maintenance contract for one additional year.

An action item (No. 31) was assigned to N. Green to develop strategies for reduction of protocol overheads for the Boston communications link by 10%. Action item 31 is due on (date).

DISTRIBUTION: M. Black, L. Blue, K. Cordovan, S. Hue

---

are attached as appendices for design team distribution but need not be widely distributed to management.

The body of the minutes should include a list of attendees, announcement of the next meeting location, date and time, a series of paragraphs summarizing salient issues covered, and a distribution list. A copy of the current operating plan schedule with updates should be attached to the minutes. Also, a list of outstanding hardware and software action items assigned at the team meetings should be attached. Separate lists of failure reports and change requests resulting from formal test exercises, or from experience with previous operational deliveries, should also be attached as required. The documents should include an identification number for each action item, failure report or change request, the name of the initiator, date of initiation, a one-line summary of the issue, the assignee for action, a due date, and a comment field. The form and format for these documents is discussed in the planning for configuration control section of this book. These reports must be integrated with the configuration control, auditing, and status accounting mechanisms as a part of configuration management.

Buzz Words

One more thought. There are a lot of buzz words and phrases out there today—words and phrases such as faster to market, flatten the organization, knock down barriers, concurrent engineering, ownership, pride, self-esteem, robust design, eliminate non–value added process, reengineering, teams, and so forth.

When any of the product development teams are working right, *they are doing all of these*. Think about it. It is at once that complicated and, incredibly, that simple.

Phase A: Product Development Feasibility

WHY PHASE A?

You're in charge. Somebody gets an idea. The idea may come from any source, from formal market research, from less formal memos, or from a bright thought generated by anyone within the organization at any level, or often from a customer, a friend, or a spouse—anybody.

Before you say, "That's great! Here's some money. Let's do it," what do you want and need to know? Would you like to see some kind of detailed development plan? Better yet, is there anything you want to know *before* committing funds to building a detailed development plan?

This is the "Tell me more about it" phase. Interestingly, you may, or may not, have to go through it in a lot of detail. For example, assume the new product is a clearly achievable low-risk technical modification of an existing product—yours or a competitor's. Assume also the market is screaming for it; the capability is right down your existing alley; you're basically all tooled up and Charlie and Harry can do the whole thing in a New York minute. Then the effort involved in documenting Phase A may be minimal. All you may need to do is look over the generic Phase A issues in the next section to satisfy yourself that a go ahead is warranted.

But we both know it's not always that simple. Many of today's bright product ideas involve lead times and considerable expenditures, and often require a multidisciplinary effort. For example, if the idea involves a new model for an automobile, or a commercial satellite, or a tooling control system using new technology, or a new airplane, or new software for the net, or anything remotely complex, there are clearly some fundamentals to be addressed before any significant commitments can be made. In any case, it makes sense to satisfy yourself that the right basic questions have been covered. But what are the basic questions?

The following discusses the generic set of considerations that should be addressed in any initial feasibility stage. This is a checklist to assist completeness. You may already routinely address some of these items; some of them may not apply to a given product development. However, the list should be reviewed to be sure you're not letting something slip by. It is easier to take an item off a list than it is to formulate a new list for each new product effort.

GENERIC PHASE A ISSUES

The generic issues to be addressed in Phase A are

1. Preliminary definition of product needs, constraints, and resources of

   end users

   your own institution

   partners

   any sponsors

2. Preliminary definition of product concepts to include

   a mission statement

   design concepts

   inherited equipment

   operations and logistics

   development and production risk assessment

   development and production resource estimates

   the development paradigm

   return on investment analysis

Remember, in Phase A these issues are addressed at a top level. Detail in one area or another may vary as required, but the detail need only be sufficient enough to enable management, upon review of the team's work, to make a sound decision as to whether to fund a detailed planning exercise or to drop the whole idea. A more complete discussion of these items follows. But, first, a few words on Phase A team membership.

WHO'S ON THE PHASE A TEAM?

Team membership is determined by what the team has to do. First, decide what issues are going to be addressed and, then, pick the personnel with the right disciplines. Follow the lead of what you want to do. Here is a candidate list:

1. The customer. If your product is for a single organization, have one or more end users from that organization on your team. For general consumer products, have someone (probably from marketing) on the team who will faithfully act as the customer advocate.

2. The institution. Get someone on the team who understands your institutional goals and its relationships with partners and sponsors, including supplier capabilities.

3. A lead hardware engineer who has sufficient understanding to articulate hardware issues or can accept action items to effectively address them with his colleagues.

4. A lead software engineer who has sufficient understanding to articulate software issues or can accept action items to effectively address them with his colleagues.

5. Someone who understands logistic support both for development and for operations. The items are the same, but the content is different for each. Logistics items include support equipment, test equipment, transportation and handling, technical data packages, personnel and training, facilities, and maintenance planning.

6. Risk identification and analysis expertise for both development and production.

7. An expert in top-level cost analysis using techniques of analogy and parametric analysis, or other viable magic.

There are seven to nine people right there. You may, of course, get away with fewer, depending on product complexity and the number of generic items you choose to cover, but you don't need more. Now, further detail on the generic issues.

NEEDS, CONSTRAINTS, AND RESOURCES

We first assess the needs, constraints, and resources of the ultimate system user(s), the developing institution itself, any partners, and the sponsors of system development. The system user is unquestionably the first and most important of these to be considered.

User Needs, Constraints, Resources

A thorough knowledge of the potential user and user environment *should always* be a part of preproject planning. We are, however, still in a feasibility study phase. The analysis of user needs during this pre-project planning stage is typically carried out with considerably less detail than the more formal user needs assessment performed as a first step in the detailed development planning phase. Complete and formal user needs assessment comes later and may include the identification of quality drivers through quality function deployment (or similar practices). But that takes place following Phase A planning, assuming that approval for detailed planning (Phase B) is granted. Even so, an accurate identification of user needs at this stage, while less formal, is still of crucial importance.

Ideas that reflect user needs for new systems, or for improved existing systems, at the Phase A stage may come from almost any source. Typical sources are:

1. previous, or ongoing, market research activities
2. on-the-job experience of employees
3. feedback from the user community
4. professional meetings, seminars, or associations
5. outside publications
6. competitor activities
7. extensions of technological trends
8. opportunities for technology integration

Any and all ideas should be oriented toward the constant quest for improved customer satisfaction.

Top-level user constraints to the implementation of new ideas should also be reviewed at this point. System designs are often constrained by existing operational and support systems with which they need to interface. Rarely is a brand new system built from scratch that is not impacted by some other system. One common user constraint is the need to use existing equipment, commonly at specific system interfaces or operational support facilities.

There are always cost and schedule constraints on any project. In some cases, however, they are absolute drivers. Examples are design-to-cost projects or space-borne systems that may need to be completed in time for specific launch windows of opportunity.

Severe user constraints may also be imposed on commodities such as power, weight, size, or levels of performance, or by design inheritance. While issues of this type may blur into classification as functional requirements (which are formalized early in the development phase), an early understanding of their potential for determining system feasibility should be sought. Does this sound pedantic? I am reminded of the words of ex-Nissan executive, Toshio Nakano, who became a Ford dealer in Tokyo, "The Taurus is a very fine car. We all like it. Unfortunately, it is too big to fit in the automated garages in Tokyo."

There may also be political, intellectual, or cultural constraints associated with the user setting that preclude any further considerations for system planning. Early Western missionaries in Australia virtually destroyed an entire native society by simply introducing an axe into their culture. It's an interesting story. Axes were crudely made by hand and owned solely by men. When young men reached puberty, they received ritual instruction as to how their individual fine axes were to be made and carried with honor throughout their lives. Women did not have axes. The axe was an embedded cultural symbol of a male-dominated civilization. When missionaries arrived and saw the natives struggling with their crude tools, they decided to help by introducing fine steel axes manufactured in Europe. Following services one Christmas morning they distributed the superior well-honed tools to all who attended, virtually all of whom were women and children. (The men did not typically frequent the churches.)

By afternoon chaos abounded; it did not take any longer. Family structures disintegrated. Soon the culture and all of its commerce rapidly degraded, as well as the trading chain up and down the river to the coast. Is the example outdated? Certainly the axe is a fairly simple product. Still, we do well to make a distinction between the subject matter and the principle of what happened. Even today, our technologies are easier to change than our cultures. We may have solutions to needs, but their successful implementations may often have to wait for the right political, legal, moral, or cultural setting.

A fine example of product sensitivity to culture is the successful expansion of the South Korean conglomerate, Daewoo, to Eastern Europe. Daewoo products range from banking, automobiles, bearings, ships, hotels, consumer electronics, securities, real estate, nuclear materials, and technology. A great deal of time, effort, and training took place, with emphasis on cultural differences and their impacts on characteristics of products and services. Daewoo made sure it wasn't simply dumping axes. Many trips were made back and forth in the interest of training—which included not only programs for the front-line providers of products and services, but also travel and training programs for the designers and engineers back home.

Daewoo took time to tailor its products and services to the needs and language of users. Another good example is the MacDonald's Canada system in eastern Europe and Asia. It was successful in moving at the right time with the right products, and with the right sensitivities, including the right language.

Is language important? I think so. I can tell you of one American giant that had to recall a major product from China because it used the word "always." When translated into Chinese, the English word "always" becomes *forever.* We are talking here about long after the sun burns its hydrogen, and then its helium, blows up and becomes a red giant engulfing us all, roughly five to seven billion years from now—and then on to forever after that. This means that if you say things such as, "We will always support you," you instantly lose your credibility. "Always" can't be true. What else are you telling me that can't be true? Bingo—your integrity is in question.

The French are wonderful people. They like wine with lunch—a good long lunch in a reasonably nice setting. Opening an amusement park with fast food (to get folks back on the rides sooner) and no alcohol allowed works in the United States. But the design doesn't help the ROI outside Paris.

In cases such as these, be assured that in-depth consideration of user needs, constraints, culture, and resources is of paramount importance in product development.

Finally, even in the absence of cultural barriers, the user may not have the resources, or infrastructure, to utilize the product. Without fuel, spare parts, and training, the product can't be used or maintained. While we may have tractors to sell and believe we see markets, the real needs of our target customers may simply be enough oxen, or blades, or food, or civil stability.

These considerations are generally recognized in our decisions to proceed with product implementations geared for specific markets. But they should always be left on the list of items to address. It is often too easy for us to assume that what we find useful, everybody will find useful. In practice, the lesson is not always so obvious. Engineers notoriously tend to plug their pet technologies into user environments. The public still jokes (and grumbles) about VCRs with too many features and a complexity they don't need and don't know how to use. Marketing personnel have been known to come up with solutions and then look for problems to fit them. In this respect, the axe story isn't all that old. The user is always of paramount importance. A complete understanding of the true needs of the user, the complete user environment, is essential. This is particularly true as technologically advanced nations seek new markets in Second and Third World nations.

In assessing user needs, we should always concentrate on identification of the very top-level user problem independent of our own orientations and interests. The orientation should be to identify the basic problem first, well before attempting to identify solutions. As product developers, we are of course oriented toward opportunities that are consistent with our institutional goals. Automobile companies don't build ocean liners. Still, the common approach of coming up with solutions, and then searching for problems, is not sufficiently user oriented, and is downright dangerous.

Institutional Needs, Constraints, Resources

Of course, efforts to identify specific opportunities for new, or improved, products is related to institutional goals. Planners in the consumer electronics industry do not spend time and money investigating user needs for locomotives. Still, within certain boundaries, the mission of any given institution can vary. The following Phase A questions and considerations are related to institutional needs, constraints, and resources:

1. What is the long-term mission of your organization? Is there an institutional mission statement contained in a single paragraph that can supply guidance? Is expansion into new, or related, markets a part of the business plan? Or is concentration on core capabilities a point of emphasis?

2. Is expansion of current capabilities and resources through selected acquisition a part of the institutional mission plan? Does the product under consideration fit into this strategy?

3. Are the necessary logistics support mechanisms in place for a product development program within your organization? Logistics support elements consist of facilities, test equipment, supply support, personnel and training, technical data packages, maintenance capability, transportation, and handling. While these logistics support elements should always be a part of any product design in support of operations, they are often overlooked as requirements for product development, test, and delivery. Logistics support elements are the same in either case, but the details, of course, differ. For example, facilities needed to support product development are distinctly different from

those that support the operational environment. The former are a part of project planning and the latter are a part of the mission product design.

In some institutions, two sales have to be made before a new product development can take place—one to the customer and the other to your own institution. Clearly, the decision to proceed with detailed product implementation planning requires an understanding of the institution's willingness, commitment, and ability to provide complete development support within and across institutional boundaries, *and to empower the development team to succeed.*

Partners Needs, Constraints, Resources

Partners in product development consist of those both internal and external. Internal partners are other departments or divisions within an organization that provide support in areas of engineering, quality assurance, review, production, management, test, and other programmatic areas, for example. Departmental barriers are common in American organizations and at all levels of governments. The modern quality movement has had some success in alleviating these barriers through concurrent engineering practices and by setting a tone for internal cooperation instead of "defense of isolated turf." The support of top management for any product development is necessary, but may not always be sufficient. Internal institutional support, cooperation, and commitment at all working levels should be sought out and solidified during Phase A.

Examples of external partners are suppliers and codevelopers. The ability of suppliers to meet the demands of both development and operational activities for the product must clearly be established. An increasing number of organizations are forging intimate and long-term relationships with suppliers as well as reducing the number of suppliers to a select few. This is a good idea. Suppliers are increasingly becoming involved as team members. They are often involved with complete subsystem or component design as a part of a concentrated effort to improve quality by efficiently integrating their products into their customers' product designs. Phase A should include strategies for supplier selection and verification of the existence of a quality-oriented, responsive supplier community.

External partners often consist of codevelopers, major subcontractors, and/or international partners. The institutional needs, constraints, and resources of these partners are as important as those of the primary developing organization. Strong systems engineering leadership is of particular importance in these cooperative situations. External partners must understand and accept that the primary developer's systems engineering office is the final arbiter of all technical issues. While both technical and programmatic lines of responsibility must clearly be established, it is much more important in the Phase A effort that all lines of *authority* are clearly defined, understood, and agreed to. A surprising number of products are under development today involving partners without clear lines of technical authority. The bigger the institutions are, the more likely this is to happen. Don't accept responsibility for partnerships, or anything else, without clear lines of programmatic and technical authority.

Sponsor Needs, Constraints, Resources

Sponsors may also be internal or external. Internal sponsorship arises when an organization wishes to develop a product to support the institutional infrastructure—for example, the development of a stock trading display system for a brokerage house, development of an organization-wide financial accounting product, development of a hospital information system, development of a marketing information support system, or internal research and development, to name but a few.

External sponsors may consist of private foundations, domestic or foreign governments, or other prime product developers who call upon your organization to function as a developer, or subcontractor.

During Phase A, it is important to define sponsor management and technical interfaces, support capabilities, financial commitments, and other relevant issues. It is vital to understand the complete spectrum of sponsor needs and expectations as well as the sponsor's ability to commit to complete product development and delivery.

PRODUCT CONCEPTS

Once a realistic set of needs, constraints, and resources is established, the technical feasibility of product development can take place. This step involves the development of product concepts.

Top level product concepts are

1. a mission statement
2. design concepts
3. inherited equipment
4. operations and logistics
5. development and production risk assessment

6. development and production resource estimates
7. the development paradigm
8. return on investment analysis

Now let's look at these product concepts individually. Work associated with each of these items is essentially carried out in parallel with constant iteration between each. This iteration is natural and expected. While each of these is discussed in order, their interrelationships are evident.

Mission Statement

The mission statement for a proposed product is documented in a succinct top-level goal statement. There are no hard rules as to the inclusion of measurable product requirements or product operational constraints, such as will be required in the development of functional requirements during Phase B. Details of this nature may or may not be included. What is required is that sufficient information is provided such that development of the remaining elements of a complete product is clearly traceable to well defined and bounded top-level concepts.

A useful guideline for developing a mission statement is to consider the current state of affairs and, then, the desired state to be achieved. The mission is simply to get from the current state to the desired state.

The current state is usually associated with an identified set of user needs, and it generally follows that the user need is not presently being adequately met. User needs may be unfulfilled because there is no product in existence to meet the need, or, because similar products exist but, for one reason or another, do not fully meet the existing set of needs. "Needs" can also include those for products that are not yet recognized by the user as necessary, or convenient. In these instances, markets are "created." But the purpose and concepts outlined here are the same. In any case, the current state can be succinctly expressed as the set of user needs. These needs may be expressed as global factors such as lack of performance, complexity of maintenance, acquisition or operational costs, failure to meet demand, a need to expand our knowledge, the whole range of esthetics, and so on. If products exist to meet the need but are simply not in place, then product development may not be necessary and procurement of an existing product with minor anticipated adaptation might be sufficient.

Mission statements should not be lengthy. For example, suppose a user needs assessment had been carried out through market research that revealed a specific opportunity in the automobile market. Assume that the research found a growing population of consumers over 55 years of age who would be attracted by a sporty looking but comfortable vehicle. The potential market is characterized by people who's kids have moved on, are on the verge of retirement years, and are likely to be on a limited income. They also foresee some traveling, they like the youthful look, but they desire comfort and easily accessible luggage space. They are not rapid accelerators or speeders.

Consider the following candidate mission statement:

Market research has identified a user need for an economical, medium-performance, comfortable, youthful, and sporty-appearing vehicle with easy access for the mature driver. No such vehicle exists in today's market. We will provide integrated concurrent engineering resources to include suppliers, dedicated internal cross-departmental assets, dealer service organizations, and continued user feedback for the timely introduction of a competitively priced automobile that is economical to operate and meets customers expectations for appearance, comfort, and performance, with a time-to-market of two and a half years.

Note that this statement not only provides a top-level characterization of the perceived need but also recognizes that meeting mission goals will require an integrated effort across many organizations and departments. In this example, an integrated concurrent approach is deemed mandatory to success and is included as a necessary part of the mission. Also note, a short lead time is included as an integral part of the mission.

Here's a completely different example. The asteroids consist of a belt of some 3,000 bodies whose orbits around the sun are largely concentrated between the planets Mars and Jupiter. Unlike the slowly changing major planets that harbor atmospheres, the asteroids are believed to be well-preserved remnants of the early formation of our solar system and to hold clues regarding processes in effect during that formation. The following mission statement is motivated by a perceived need to further our knowledge.

Our knowledge of planetary bodies has been greatly enhanced through planetary missions over the past two decades. These bodies, however, have typically gone through vast changes since the origin of our solar system. We now have the technology to carry out extensive investigation of the historically stable more primitive asteroid bodies. We will design and execute a relatively low-cost mission to characterize as a minimum the following properties of a selected set of asteroids:

1. *size, shape, and mass*
2. *surface structure*
3. *physical makeup of surface material*

The mission shall include study of the largest asteroids; it will secondarily investigate other targets of opportunity during the mission lifetime.

This mission statement example succinctly describes the perceived top-level current state of our knowledge and a specific desired state of our knowledge. It also calls for a low-cost mission relative to past larger-scale planetary missions in keeping with the national trend to economize on science. Minimum science goals are also clearly set. Finally, the statement uses the word shall (akin to a requirement) to specify that the largest asteroids will be prime targets for investigation even though specific bodies are not identified at this point. Lesser bodies will be included in the mission design as finalized trajectory opportunities permit.

It is evident that there is considerable freedom in the content of the mission statement. The statement is important, however, in that it clearly establishes top-level goals (and sometimes requirements) to which all subsequent work must adhere. Writing mission statements is not easy. The words should be well chosen because they will have direct and lasting impacts on the remaining elements that go into a completed product concept, as we shall see.

Construction of the mission statement is the first step in the development of product concepts. Development of the remaining activities can take place in parallel. Should these subsequent efforts lead to clear deficiencies in the original mission statement, then the statement should be changed to reflect this need for completeness. As a rule, however, mission statements should not be constantly altered without considerable thought and only when clear improvements are indicated. That is, while we may freewheel and brainstorm among and during the development of the remaining product concepts, we should strive to maintain the integrity of the mission statement.

Mission statements should always be approved by management and by either the user community, or a representative sample of the user community, or by marketing personnel who can faithfully represent the user community.

Development of the mission statement is not an academic exercise. Its content will be put to *important* use in later phases for development of product priorities. (Priorities are defined during the product development planning phase.) In modern systems engineering, these considerations have major uses for the consistent decomposition of requirements and in design tradeoff studies during the actual product development phase.

Design Concepts

Design concepts are developed to assess the technical feasibility of a mission statement. They help establish whether the mission statement can be met within an acceptable risk envelope, or must simply be rejected. Design concepts at this stage do not involve a detailed design but are sufficiently engineering oriented to validate technical feasibility. This does not mean that some applied research and development may not be required in specific new areas. Needs for new R & D are also evaluated for realistic and timely accomplishment and are subject to appropriate risk analysis.

In our previous automobile mission statement example, design concepts might consider such top-level design issues as wheel base (related to comfort), number of cylinders, and consumer fuel and maintenance costs (related to economy). We also note that performance standards related to compression and acceleration may not, in themselves, be cost drivers. A new styling, however, is likely to be required (related to a youthful and sporty appearance). But features attending to interior seating, ease of entry and exit, and trunk accessibility (ease of access) may well be inheritable in a mature automotive firm. The mission statement also requires that retail cost, and operating costs also, be given constant attention. Operating costs, of course, relate not only to fuel efficiency but also to maintenance and repair costs. Thus, parts reliability and diagnostic ease are features to be emphasized in the design. Note how these pertinent design issues are logically derived from the top-level mission statement.

For the asteroid exploration mission statement, design concepts would most certainly consider spacecraft size, mass, and complexity. These issues directly relate to development costs, launch costs, and operational costs. They also impact the number, size, mass, and type of instruments to be included for the desired science. Mission trajectory analysis must also be studied to ascertain a mission course to encounter a reasonable selection of large asteroids. The encounters must have sufficient time on station to adequately complete the stated science goals. Mission life and on-board fuel tradeoffs are made at a top level. Use of a number of small spacecraft versus a single larger vehicle may also be considered.

In both of the above examples, we observe that the major design issues logically derive from the mission statement. Secondary and tertiary design issues are best identified in a logical product breakdown structure and brainstormed to lower levels of detail in order to judge feasibility of the design concept. The product breakdown structure is carried as far as required to adequately address technical feasibility at an acceptable risk level, and no further. Thus, all issues addressed in the formulation of design concepts should be directly traceable to the mission statement or logically derived at lower levels from a product breakdown structure based on the top-level mission statement.

During this exercise, be aware there is a tendency to delve into detail beyond that which is required during Phase A. I once ran a Phase A team investigating the feasibility of a high-tech communications system. The concept called for a large, lightweight switching matrix which was critical to the design. I assigned one of the team engineers to look into its feasibility. A week later, I found him looking through a parts catalog steeped in a detailed design. This highly technical professional lacked the mind-set to work at the level required in Phase A. He could not commit without knowledge of every single detail. Phase A engineering team members should be able to call upon their broad experience and be comfortable developing design concepts at the right level. Detailed design is *not* required during concept formulation. The purpose, rather, is to simply gain sufficient confidence that the goals of the mission statement can in fact be accomplished within an acceptable risk envelope.

It is also possible that technologies are insufficiently advanced to accomplish mission goals within resource estimates. Further basic research may be required before development can take place within the desired risk envelope and time frame. In this case, the mission statement must either be modified, or a realistic decision made not to initiate detailed project planning. That is, the idea might just have to be abandoned.

Inherited Equipment

Technological heritage is always involved whether we are developing a new product or modifying an existing one. Technological heritage is a broad term referring to all corporate baggage and history. Our interest here is with a small part of this "heritage"—inherited equipment. Inherited equipment consists of any tooling, machinery, parts, supplies, or other accoutrements used in prior product developments. The question is, to what extent is this equipment useful for the current project? The answer is driven at the highest level by the content of the mission statement.

The automobile mission statement example calls for a "new" market offering. If we assume the mission statement was written in the mid 1990s, the lead time is of the order of 24 to 48 months, the late 1990s. Clearly, the risk could be greatly reduced if an existing economical engine, or a moderate modification to an existing engine, could provide the needed power plant. Since the statement calls for medium performance, successful use of an existing (inherited) resource would be a likely starting point. This includes the potential use of existing tooling and suppliers. It would be desirable, however, to built a high level of quality into the engine from the standpoint of reducing mean time to failure and mean time to repair, that is, user operational costs. The vehicle will also require a bold and imaginative new body design.

Alternatively, a wider base and lower center of gravity may be considered as extensions of existing technologies. Issues related to comfort, trunk accessibility, and ease of ingress and egress do not appear to be technologically threatening.

Inheritance issues thus wend their way into design concepts by assisting in the division of needed features into two categories: those that may be implemented through modifications to inherited resources and those that will require substantially new designs.

A similar top-level thought process may be applied to the spacecraft example. Review of existing spacecraft products and internal products such as attitude control, propulsion, navigation, and so on, for applicability to the "new" mission would clearly be in order. Because the particular science called for has been similarly accomplished in the past, existing instruments—or modifications to existing instrument technologies—may be of significant use. Modifications of particular interest would involve smaller and lighter components. What is new in this mission is the decidedly different and potentially multiple trajectories for the spacecraft, the smaller than traditional size of the spacecraft, and the potential use of multiple platforms to achieve a complex set of rendezvous. Again, the inheritance issue focuses on how the use of inherited resources versus the need for new capabilities impacts the complete set of design concepts.

Operations and Logistics

Product design concepts should be heavily influenced by how, by whom, and where the product is to be used. While this seems a simple reality, the operational setting is often not given sufficient attention during product development. In some national, state, and local government system procurements, funding for development and operations come from two different sources so that life cycle costs, for example, are never really considered during development. This prevalent and serious problem is almost always associated with bureaucratic organizational structures.

During the development of product concepts, the operations concept is intended to address the following considerations:

1. a description of the operations organization
2. a top-level description of functions to be performed during operations
3. a top-level description of interfaces of the operations organization to other organizations
4. a top-level description of logistics support required by operations which include maintenance, personnel and training, transportation and handling, supply, facilities, technical data packages, and test equipment

The details of these ingredients, of course, differ widely, depending on the product in question. For automobiles in use, the operations organization consists of at least the complete set of service facilities. The consumer, or user, may be considered as a part of this organization or as an interface to it. Because economy is an important part of the mission statement in our automobile example, the operations concept must also become an integral part of the low cost goal. Thus, the design concept must directly address this operational need by providing for ease in fault isolation and restoration of operational capability in a reasonable time frame and at a reasonable cost. This concept can be further broken down into concepts of on-board diagnostics, modularity, parts reduction, dealer maintenance training, test equipment, involvement of suppliers in quality, and so on. This is true not only for repair facilities under direct control of the manufacturer, but also for independent service organizations. Car owners do go to independent repair facilities. That is, our attention is always focused on the customer's best interest. Clearly, the call in the mission statement for integrated concurrent engineering is well thought out.

Our spacecraft example requires an entirely different operational organization. The four points outlined above, however, still hold. Basic operational functions include spacecraft launch, tracking, engineering, navigation, cruise and encounter science operations, and the distribution and archiving of scientific data. For example, tracking operations can have a major impact on spacecraft communications system design. Principal interfaces are with other ground support products, suppliers, contractors, the public information office, the sponsor, and the broad science community eagerly awaiting results.

Impacts of operational concepts on design concepts are of extreme importance for one fundamental reason. Not only does operations strongly influence design, but the operational environment is where the customer lives. The primary reason we develop products must be to satisfy customers. All other reasons, no matter what importance we may wish to ascribe to them, are secondary. If you are commercially oriented, that includes making money. Of course you must still sell units for more than they cost, but the profit margin is not what makes you rich; it's happy, repeating customers.

Risk Assessment

Few undertakings are free of risk. Risk management is an organized means for identifying what can go wrong, analyzing potential impacts, devising alternative strategies, and executing those strategies to remove or alleviate contingency as it arises. Risks can, and usually do, arise in schedules, requirements development, funding profiles, actual costs incurred, design, and virtually all technical areas. Risk identification must be an up-front issue. Risk management is a continuing effort through all phases of product development.

In Phase A, risk management is focused on the identification and top-level assessment of risk associated with each of the other ingredients of the product concept. These include the design concept, use of inherited equipment, operations concepts, resource estimates, and the development paradigm selection. In this preliminary stage, it is useful to identify all potential risk items and then to rank them in terms of low, medium, or high. A conservative approach is to list as many items as may occur to the preproject team in each of the product concept areas through a brainstorming session. For each risk item identified, a succinct statement is next written down covering each of the following topics:

1. a statement of the potential problem
2. estimates of the consequences of failure in terms of cost, schedule and/or performance
3. an alternative backup strategy, or strategies, to remove or control the risk
4. conditions, tolerance thresholds, and rough timing under which each strategy would be implemented

These four points basically enable the Phase A team to say

▰▰▰▰
••••••••••••

We believe we have identified major risk issues, we understand their consequences, we have a strategy to handle each issue, and we have a time frame to implement that strategy to minimize risk to the budget, the schedule, and the quality of the product."

••••••••••••
▰▰▰▰

That's nice.

The fact that risk exists does not necessarily need to be a frightening thing as long as sensible backup alternatives and their timely use can be established in advance. Risk can be greatly diminished by devoting sufficient thought to items 3 and 4 on page 53. (Note: with regard to item 4, one way to dramatically reduce and control risk is to use the rapid development model for product development. More on this in the next section.)

Should approval for detailed planning occur, a complete risk management plan will be developed during project planning and executed by systems engineering during implementation. At this point, however, we are identifying major risk issues and top-level strategies for their reduction to an acceptable level in the interest of validating or rejecting our product concepts.

Resource Estimates

Resource estimates consist of predictions for in-house workforces, a schedule time line, and costs to be incurred for detailed project planning, complete product implementation, production, and marketing.

Costs can be grouped in four categories: those for in-house workforce, services, procurements, and overhead. Services include such items as maintenance, publication services, accounting, and so on. Procurements include subcontractors, consultants, travel expenses, project-specific equipment and supplies to directly support planning, implementation engineering, production, marketing, logistics support, and testing. Overhead includes costs for line management, secretarial support, nonspecific project supplies and equipment, benefits, and so on.

Phase A takes place without the benefit of a detailed signed-off work breakdown structure (WBS). Without a detailed WBS, bottoms-up scheduling and costing cannot take place. Bottoms-up scheduling and costing consists of constructing work precedence charts for the lowest responsible WBS accounts, placing time lines on the work, resolving discrepancies, loading the schedules with costs, and the upwards integration and iteration of these data to assemble and resolve higher-level schedule and cost constraints. Bottoms-up resource planning takes place during detailed project planning, Phase B. In the preproject Phase A, we do not employ, or need, the resources to carry out such detailed scheduling and costing.

Alternative and more appropriate techniques commonly used for top-level costing estimates include parametric analysis and the use of analogy.

Parametric analysis consists of the use of broad existing databases gathered from hardware and software experience on similar projects. The data are correlated to a minimal set of characteristics such as size, acceleration, power, weight, and so on. The computer-based model PRICE (Programmed Review of Information for Costing and Evaluation) is one widely used resource among military, aerospace, and commercial organizations for this type of analysis.

Analogy is the best approach in Phase A because it is based on direct experience. Analogy simply relates functional, physical, and performance characteristics to similar known existing items. Modifications are made to account for differences in configuration. Schedule estimates are also best made from analogy with appropriate modifications for identified risk areas.

These early resource estimates, while not complete, are of fundamental importance in establishing feasibility. Such resource estimates either fall within, near the boundary of, or outside of recognized limits. The uncertainty also provides an initial indication of the amount of programmatic margin that project management may require to carry out development.

Choosing a Development Process Paradigm

This one is very important. Of all the topics covered so far, this may be the most unfamiliar. It may even take a little shift in mindset.

It turns out that there is more than one fundamental way to approach the management and implementation of product development. There are at least four—and even clever combinations of them. These approaches, or paradigms, are very interesting because if you pick the right one at the outset, the probability of avoiding programmatic overruns is greatly enhanced. But if you pick the wrong one, you can easily get in real trouble and never know why. Oh, you'll come up with reasons, but you will never really know why. What is even more interesting is that if you pick the wrong paradigm and fail, and have the resources to try again, you will be pushed toward the right one. It is far better to see all this up front. Hence, the subject deserves some attention.

Over the past forty years, process paradigms for product development have evolved through the staircase, early prototype, spiral, and rapid development method models. The evolution is characterized by successively deeper penetrations of prototyping concepts into the traditional staircase paradigm. Early prototyping penetrates to the requirements stage. The spiral model penetrates to the requirements stage and beyond to the design stage. The rapid development method, the newest innovation, entails the repetition of the entire staircase process, resulting in repeated, and increasingly sophisticated, *operational* deliveries. This is not to be confused with preplanned product upgrades, which are merely additions under one of the other three paradigms. The difference has to do with an understanding of requirements—that is, whether they are completely known at the outset, or they are partially unknown. More to come on this.

Each paradigm continues to have its useful niche, but selection of the wrong model for a given task invariably increases the probability of failure. An awareness of the origins and use of these options is of extreme importance in selecting the right paradigm for any product development process. The selection should be carefully considered and justified at this point in the Phase A planning process.

The emergence of the newer development paradigms has not been widely described. A brief overview of this history is of interest.

Build It, Test It, Fix It
Initially, the common paradigm was "Build It, Test It, Fix It." The industrial revolution and the advent of mass production began to change that. Some industries took a while longer. For example, during the initial growth of access to computers in the 1940s this was, in fact, what software engineering was mostly about. For smaller products, the approach worked quite well. It may still be an acceptable approach when the developer is also very close to the user and there is little need for external communication. For example, it is common in materials science for a customer to approach a potential provider with exact specifications for an improved product. If the materials company management feels confident that the challenge is right down its alley, management may simply set an upper limit on funds and timing for what is essentially an applied research activity. It then cuts the brightest one or two people loose to get the job done. And that's the extent of Phase A, Phase B, and product development.

Historically, however, with product developments that required multiple disciplines and longer time frames, it soon became evident that this approach often compromised adherence to user requirements, adequate testing, acceptable documentation, and maintainability. It also resulted in a horde of other shortcomings. However, the approach typically generated products that, when finally "fixed," tended to exhibit ungraceful architectures in the form of appendages upon appendages.

The Staircase Model

With the advent of 'big' products, better ideas emerged. The building of the SAGE (Semi-Automated Ground Environment) product at TRW in the mid 1950s was among the first recognized attempts to do business in a more structured manner. The philosophy was to execute discreet and controlled stages of product development that consisted of the orderly determination of requirements, specifications, and design, followed by fabrication, test, and the performance of acceptance testing prior to operations.

The approach was soon expanded to recognize that one needed to at least partially explore a subsequent stage in order to gain full understanding of the present stage. The modified model visualized a block of effort that moved forward in time, encompassing the present stage and the next stage (see Figure 4.1). While this paradigm upgrade did not recognize a need for feedback across multiple stages, it clearly helped to introduce a reality of human behavior (our ability to fathom complexity) into the process. *This remains the way just about everybody thinks today!*

That's right. The staircase paradigm has had a powerful and lasting influence on the way we develop products and do systems engineering. It gave birth to an entire lore of control mechanisms to ensure the maintenance of its orderly flow. These new disciplines and techniques included classic configuration management, standardization of reviews, formalization of the product design team, and third-party quality control and test teams, to mention but a few.

In time, it also gave rise to a host of ancillary techniques designed to maintain the integrity of the staircase concept when it more than occasionally struggled or faltered. These additional tools included structured analysis, structured design, structured programming, object-oriented programming, program design languages, unit development folders, computer-aided design tools, and other similar techniques designed to maintain control and visibility. Throughout this period, the staircase paradigm remained solidly entrenched. The belief

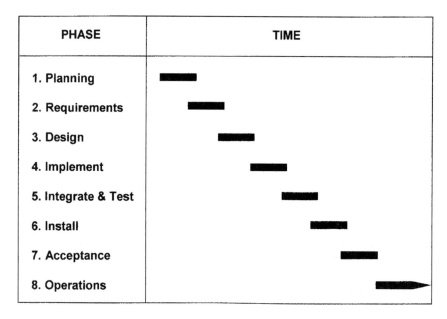

PHASE	TIME
1. Planning	
2. Requirements	
3. Design	
4. Implement	
5. Integrate & Test	
6. Install	
7. Acceptance	
8. Operations	

Figure 4.1. Traditional Waterfall Schedule.

persisted that the model could inherently be successful when accompanied by adequate controls and sound management.

The staircase model remains valid, if certain conditions are met. Perhaps the most important condition for its success is that customer requirements are completely understood at the beginning and can be stated in clear and measurable terms. Use of the staircase model also requires mature technologies with minimum risk. The next step in evolution, early prototyping, found its impetus through recognition of the fact that this is not always the case.

Early Prototyping
Some industries, such as the automobile and chemical industries, have been routinely prototyping and testing for a long time. But it was not until the late 70s and early 80s that software-intensive products began to formalize the idea. A Government Accounting Office survey, reported in October of 1980, disclosed that only 2 percent of the products surveyed were used as delivered.[1] Fully 47 percent were delivered but never used, and 29 percent were never delivered. The remaining 22 percent were used only after moderate to extensive rework. Cost and schedule overruns accompanied many of these failures.

Evidently, users were simply not getting what they anticipated. The problem was perceived to be largely due to a basic failure to accurately define true requirements. A realization emerged that there are at least two conditions under which the user is unable to state requirements in sufficient detail to realize success. The first occurs when users cannot state exact requirements but can recognize them when they are exposed to them. The second condition arises when no one really knows what the requirements are in detail until the product has gone through a number of actual operational iterations. An example of the first condition might be a new end-user command and control product for which the user may not know the exact desired information content or the form and format for screen presentation in advance. An example of the second condition is the space station: What is the mission of a space station with a life of, say, twenty years going to be fifteen years from now? Not a lot was done immediately about the second problem. But the first problem was attacked throughout the 1980s by way of expanded use of early prototyping. The concept of early prototyping accepts that the user and the developer may not always know what the exact requirements are for a product. In this scheme, user requirements would be drawn out by constructing a set of "best guess" requirements at the outset. The set would then be rapidly prototyped and tried out in a mock-up of the user environment. Appropriate modifications would be made until the user, or the developer, was able to clearly identify the desired product behavior. When requirements were finally determined in this manner, the product development paradigm could then revert to the staircase with feedback model for all the remaining stages.

In the software world, the concept of early prototyping was widely extolled and spread rapidly and comfortably into the process of product development. Not only did individual projects embrace the new paradigm, but soon whole laboratories devoted to early prototyping appeared as organizational resources that projects could call upon to assist in the accurate determination of fuzzy requirements.

Despite these advances, in 1987 the Defense Science Board Task Force Report on Military Software concluded that traditional software development paradigms were actually hindering effective software development.[2] The Board was not alone in it's observations. The same thing was happening in commercial software intensive product development. Significant product failures were still being observed and software engineers were responding. Among the responses was a new and significant evolutionary concept in software products engineering which began to appear in the literature in the mid to later 1980s.[3] It was called the Spiral Model.

The Spiral Model

The spiral model extends the concept of a single early prototype to a series of three prototypes. The paradigm encompasses multiple iterations between requirements and design concepts, with successive refinements. The process is commonly depicted as a spiral converging on a final design. It then reverts to the staircase paradigm upon the final establishment of a design.

The first prototype aims at developing product and operational requirements, the second aims at developing software specifications, and the third prototype results in the establishment of the final design. The spiral model may incorporate more or less than three prototypes, depending upon the need. Each spiral involves extensive risk analysis, the use of simulations, models and/or benchmarks, and user interaction as required. While the impetus for this evolution came from software products engineering, the concept is clearly applicable to the larger arena of product development in general. The approach basically recognizes that uncertainty typically exists beyond the initial requirements stage and deals with this issue by reaching deeper into the conventional model with a series of penetrating prototypes.

The spiral model exhibits a very distinct and important advantage over the earlier paradigms in that it provides for unforeseen change deeper into the process. It provides a mechanism for dealing with uncertainty and the capacity to accommodate tradeoffs between implementation and requirements. In the spiral model, requirements are not really frozen until design, cost, and risk issues are adequately addressed. Further, the older paradigms tended to separate the user from the developer early in the process, each to go their merry way until perhaps a not so merry delivery. Clearly, one lesson to be learned from the observed progress in paradigms is that if the user, or good user representation, is an intimate partner in trade off analyses, the probability of success is greatly enhanced. The tendency for engineers to design products for engineers as opposed to designing them for end users is thereby somewhat reduced.

Finally, proponents of the spiral model suggest that in conditions where uncertainty is not so threatening, the approach can revert at anytime to the staircase with feedback. In this sense, the earlier models can be viewed as subsets of the spiral model.

Use of the model, however, does require access to experienced risk assessment personnel as well as experienced managers. It also requires a further modification to our classical ideas about configuration management. Use of the model does away with conventional concepts of frozen baselines at the requirements, design, and implementation levels. Unfortunately, there are old timers who absolutely insist on adherence to this idea. In the spiral model, the conventional review process involving user requirements reviews, preliminary design reviews, and so forth must now be modified to recognize the volatility of solution choices and their potential impacts on requirements and design throughout the spiraling process. The spiral paradigm requires good management skills.

The Rapid Development Model

Although the spiral model represents a significant advance in the evolution of product development and systems engineering, a further evolutionary concept designed to deal with the issue of obscure requirements has begun to emerge. This newest concept addresses the issue: What happens when a substantial portion of true requirements are not really understood until the product is actually operational? When this occurs under the conventional development paradigms, the resulting product is typically discarded because the required changes either outstrip the ability of the design to easily respond to the newly discovered requirements, or the prospect of reinitiating a new staircase fix (i.e., starting all over again) is simply too costly or time consuming.

Clearly, such products exist—more than we may wish. In the early 1990s, a new and imaginative approach, called the rapid development model (RDM), emerged.[4] The RDM was designed to extend the prototype trend to include repetitive iterations of the entire traditional end-to-end product development process. Because it is relatively new and not widely understood, the RDM concept deserves supplementary discussion.

The RDM process differs from traditional product development processes in at least two major ways. These can be summarized as

- the use of incremental product deliveries to the operational environment through repetitive use of the staircase paradigm
- use of an evolutionary approach emphasizing extensive user involvement, formal feedback from products to requirements, and incremental formality

Figure 4.1 shows the traditional schedule for the staircase product development process. Eight phases of development are shown to include planning, development of requirements, design, implementation, product integration and testing, product installation/delivery, certification or acceptance (as required), and entry into the operations

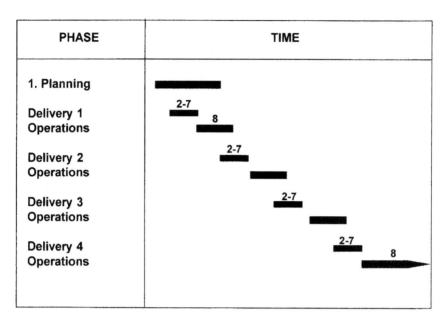

Figure 4.2. RDM Schedule Concept.

and maintenance phase. The staircase process is essentially a serial one with appropriate feedback among individual phases.

Figure 4.2 exhibits the phased delivery concept in a typical RDM schedule. Initial planning includes the careful definition of the number of deliveries to be made, and the content and timing for each delivery. Following initial planning, steps 2 through 7 shown in Figure 4.1 are duplicated for a number of iterations. Each iteration, however, has practical operational use indicated by step 8 following each incremental delivery. The effort involved in each delivery encompasses the entire staircase process, and each delivery represents a further hierarchical accomplishment on the road to final detail. Requirements for delivery N are finalized based on experience gained in development of delivery N-1.

Figure 4.3 presents a typical RDM schedule in more detail. It is generally accepted that the delivery intervals should be fairly constant throughout a complete rapid development implementation in order to maintain the desirable characteristics of the concept. The schedule example of Figure 4.3 shows a 12-month delivery cycle for increasingly complex operational products.

Each delivery cycle begins with development of improved definitions for both the remaining incremental deliveries and the final

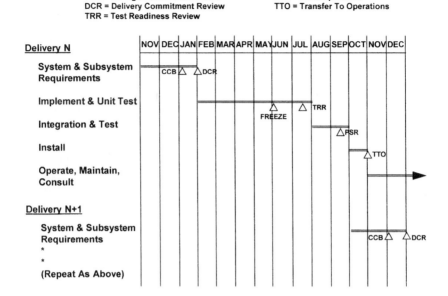

CCB = Configuration Control Board	PSR = Pre-Ship Review
DCR = Delivery Commitment Review	TTO = Transfer To Operations
TRR = Test Readiness Review	

Figure 4.3. RDM Schedule Example.

delivery. Conventional configuration management practices are modified to accommodate the model. A change control board (CCB) with customer representation approves requirements, and a firm delivery commitment review (DCR) takes place. The CCB firmly establishes realistic requirements for that delivery. Implementation and unit test follow, with an appropriate design freeze and test readiness review (TRR). Delivery product integration and test is then completed with a pre-ship review (PSR). Installation concludes with a transfer to operations (TTO). All reviews are conducted by the CCB. The delivered product is then used by the customer. Feedback is obtained for input to subsequent delivery requirements.

In this manner, requirements and designs for the next delivery are well defined, complete, and controlled. The precise definition of each subsequent delivery remains less complete until its actual staircase cycle takes place. With the completion of each cycle, however, deliveries become better defined. This evolution naturally involves intensive user involvement and greatly enhances the probability of meeting true customer needs.

Use of the word "rapid" does not mean that the complete product is necessarily developed rapidly. Rather it means that the final product is brought to fruition through a series of deliveries, each of which is, typically, some ten to eighteen months apart. The effort involved in each delivery encompasses the entire staircase process, and each delivery represents a further hierarchical accomplishment on the road to final definition. These mini-waterfalls, or staircases, allow successive evolutionary experience in the actual operational environment and in so doing absolutely ensure continual and intense interaction between developers and users. This important aspect of the rapid development model is consistent with the tenets of the modern quality movement.[5]

Deliveries are product oriented and should provide specific useful capabilities that represent incremental advancements toward the final goal. Useful guidelines for incremental product definition are that each delivery should be

- product oriented for accountability
- separately implementable
- delivered and operated as a functional unit
- comprised of simple interfaces
- of similar levels of complexity
- comprised of logically related functions
- documented as a unit for the user and maintainer

Throughout this development scenario, detailed functional and design requirements for subsequent deliveries are free to evolve in each phase based on experience gained from the previous delivery. This involves a decidedly different mind-set from the staircase paradigm. In the staircase model, requirements must be frozen prior to any implementation; whereas, final product requirements are not actually known using the rapid development method until the last phase begins. It is also possible under the RDM that requirements perceived at the beginning may be altered or even abandoned as improved insight toward the target product is gained. In the end, however, requirements are almost guaranteed to be correct.

The freedom, however, is not total. The functional capability of each delivery should be well defined during the initial planning phase for each delivery.

Is this all real, you ask?

Let's try some examples.

Suppose we wish to automate an in-house manual just-in-time (JIT) system in support of a product assembly facility. Of course, the first thing we need is an acronym: let's call it JAMS for just-in-time assembly management system.

The JAMS is a product for which the institution itself is both the sponsor and user. Automation of a manual process typically alters the subject process. In fact, it often provides an opportunity for process improvements. So we recognize that even though we already have some manual JIT functions in place, the mere fact that we are going to automate the process means that there are likely to be expanded uses of the system that we cannot pinpoint exactly until we actually begin using it. An expanded JAMS may include ancillary uses for billing, accounting, inventory management, management reporting, and other unforeseen applications. How about support for product planning, costing, assembly floor construction, personnel assignments, training, quality control, and others?

Use of the staircase paradigm requires an understanding of total system requirements at the outset, and works great when that understanding is present. But that is not the case in our example. Early prototyping and the spiral model, while delaying the commitment to finalize requirements, still call for complete system implementation prior to actual operational use. In our example, there are real questions as to how the system might evolve based on experience with future use. We would like to be sure that the initial architecture, including interfaces, will allow expansion in as yet unknown directions. We would, at the same time, also like to get something running without going through lengthy major hoops to develop uncertain future requirements that are vague and unknown to us at the outset.

What do we do when these kinds of doubts arise? We use RDM. Here's a Phase A approach for our example:

Figure 4.4 provides an early concept for the JAMS top-level architecture. There is a central core program that handles all algorithms. Requests come in, orders are generated, equipment is received, and inventory is updated. At some later time, we add billing and accounting capability, management reports, and other capabilities unknown at the outset.

Table 4.1 shows our initial concepts for a sequence of system deliveries. Now here is the important point. At the outset of the development, we only need develop requirements for System 1. These are largely known from experience with the manual system. Requirements for System 2 are somewhat known, but we shall not develop them in detail until late in the System 1 development process. Requirements for Systems 3, 4, and 5 are progressively more vague as we begin System 1. This is OK. It is even desirable. Don't worry,

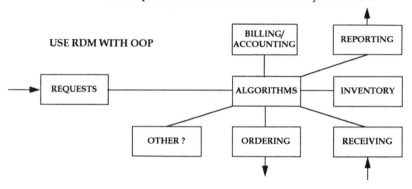

<u>JUST-IN-TIME ASSEMBLY MANAGEMENT SYSTEM (JAMS)</u>

OBSERVATIONS: we wish to automate present system which is totally manual,

new system will have capabilities beyond the present system,

present manual system requirements are understood, but

future capabilities and interfaces are not clearly understood.

USE RDM WITH OOP

Figure 4.4.   RDM Example.

❘❙ TABLE 4.1 *Sample RDM Development Approach for Jams*

System	Properties
1	Use 1 PC, develop request, inventory, ordering, and receiving algorithms. Use existing formats with some modifications as desired.
2	Develop initial algorithms for system control, billing/accounting interfaces, and management reporting. Build upgrades based on System 1 feedback.
3	Select configuration (single CPU, network, internet with suppliers?). Build upgrades based on System 2 feedback.
4	Build additional algorithms (product planning, costing, assembly floor construction, personnel assignments, training, quality control, and others?)
5	Upgrades based on experience, apply to other institutional JITs?

requirements will emerge in accordance with our phased schedule. Remember, each iteration in the RDM model calls for a repetition of the staircase model, in which requirements for each progressive system are, in fact, frozen.

Our Phase A team engineers recommend that object-oriented programming be used because we may want to be able to grab basic objects in the future to perform as yet undefined tasks. Further, they recommend C++ as a language because of its significant power and efficiency. Notice the strong hint of risk control in all of this. We will deliver System 1 in a six-month time frame following the generic schedule example given in Figure 4.3. This is exceptionally nice because we will get something useful up and running in a short time frame without carrying out a 30- to 40-month, full-blown project, and the probability of meeting requirements in the end is greatly enhanced!

Be advised, however, that the lore of product development (and that of systems engineering) is not entirely used to this. It takes good, competent, rational, and willing managers who are not locked into convention. And it requires strong leadership.

Here's another real example we are all familiar with. Consider the American space station, which began in the United States with the traditional staircase paradigm. Accordingly, all requirements were to be developed in advance. But how does one develop requirements for living space, life-support systems, and science and laboratories for a mission life of ten to twenty years? You don't. If you adopt the RDM approach, you might do something along the following lines:

Product 1	Living space module for a crew of two, life-support module for a crew of two for one month, with international docking capability.
Product 2	Modify and improve crew and life support modules, add science/laboratory module.
Product 3	Add second crew and life-support modules to accommodate four crew members for six months; modify and improve science/laboratory module.
Product 4	Modify and improve all module capabilities; add crew and life-support product modules to accommodate crew of eight for one year; add second science laboratory module.

A bit late to do all this now, but you get the point. That is, start such a project by developing definable detailed requirements for

Product 1, build it, and fly it. Develop later product requirements when we're smart enough to.

But that's not what the space station program did. We used the old staircase paradigm. In practice, our space station requirements were never really finalized prior to the beginning of design activity. There is still lively discussion regarding just what the requirements really are. Schedule and cost overruns resulted in rescoping and downscoping a number of times. We actually went through five space station "staircase" redesigns in nine years. No product.

Recognition of the rapid development method, with it's iterations of build-deliver-learn, and then do it better, would have been much preferred. The staircase paradigm fell apart in this application. The iterations and scale-backs of the U.S space station carried out in an effort to get program control were, in fact, an unrecognized shift toward the rapid development method. The "law" forced us to it. What law? The following law:

If you pick the wrong paradigm, and you have enough resources to keep trying after you fail, your efforts will migrate toward the right one at great expense of time and resources.

Meanwhile, the step-by-step Russian MIR is still flying. It was designed to employ phased and increasingly sophisticated operational capabilities, with lots of deliveries along the way. Finally, we joined them. It was a smart idea.

One more real example. In the early 1990s, I was enlisted to try to do something about a resource allocation software product that had been under development for three years, under two different engineers, and had gotten basically nowhere. The manual scheduling system was to be automated. The first thing I did was ask the manager who approached me if I would have the *authority* to solve the problem, as well as the responsibility. He said yes. Having established this, I reviewed the existing product requirements. Then I formed a team of the key people and at the very first meeting said something about the need for better requirements. The response around the room was unanimous. "Oh, no, not again." It was instantly clear that they had been worn out by trying to come up with requirements. The requirements they had under conventional systems engineering leadership were bad; they were bad because no one could articulate them. The people who had been manually making schedules, and adjustments in schedules, for resources were incredibly proficient. The process was so complex that they could not really explain how it was done, let alone guess what additional uses might be made of an automated system. All previous attempts followed the staircase paradigm. Get those requirements frozen up front. On day one I saw the development paradigm we needed. It was RDM. In six months we had a system up and running that was usable. After three more complete RDM cycles over the next year and a half, we had a complete product in place. The last product supported a meeting room environment, communications capability, and a training module. The team thought I was a hero. They received formal award recognition from top management all the way across the country. All we did was use the right paradigm.

We now know, or course, that the overall pattern of the evolution was not a planned one. The attempts of product planners to devise schemes to avoid failure did not take place with the benefit of an overall vision or direction. Rather, they were initiated by creative individuals attacking, step-by-step, what was perceived to be the most pressing contributor to the continuing occurrence of failures. I'm not really that smart, the pattern is seen only in retrospect.

There is a very important reason why the newer paradigms have been used with some success in selected situations, whereas the older ones have failed. The reason is that the newer models tend to benefit from user feedback further into the implementation process, while the older paradigms inherently assume that the implementors can essentially be cut lose in quasi-isolation as soon as the requirements baseline is frozen. The latter strategy does not always work. It is quite likely that any successes realized by use of the newer development models is solely due to this reality. Perhaps the most important observation to be made from all of the history noted is that the user, in any paradigm, must be consistently and intimately represented throughout the entire design and implementation process. It is also clear that management skills associated with the more recent paradigms are more innovative and demanding.

Development Paradigm Summary

Table 4.2 summarizes the historic motivations and approximate time frames for the evolution of each of the paradigms discussed.

Table 4.3 lists a number of application domain characteristics and examples for each of these paradigms. The examples are not absolute but are typical of their categories. Planning for use of the right paradigm should be integrated with all aspects of developing product concepts.

❚❘ TABLE 4.2	The Evolution of Product Development Paradigms
Paradigm	**Motivation/Time Frame**
Build, Test, Fix It	Lack of experience during the 1940s during the rise of new computer based products. The approach compromised requirements, testing, documentation, and maintainability.
Staircase	Realization that specific phasing was required consisting of controlled development of requirements, specifications, design, fabrication, test, and integration. Characterized by the TRW development of the SAGE product in the 1950s.
Staircase with Feedback	Recognition in the 1960s and 1970s that phases could not be isolated—that feedback between each successive phase could greatly enhance success.
Early Prototype	Recognition in the late 1970s and early 1980s that the customer did not always know what was needed. Early prototyping developed to define requirements.
Spiral	Recognition during the 1980s that implementation options were not always clear at the outset. The spiral model, developed at TRW, extended prototyping to the design phase and added risk analysis.
Rapid Development	Recognition in the late 1980s that required product behavior could not always be determined until some operational experience took place. JPL formalizes the repeated product delivery concept.

The Preliminary ROI

Expectations for return on investment vary greatly from industry to industry and product to product. Big-ticket aerospace products may exhibit ROIs of as little as 4 percent in a given year, while highly successful semiconductor and software companies have shown ROIs in the 30 to 48 percent range.

ROI estimates involve at least two components: (1) the date upon which cash flow becomes positive, and (2) the rate of growth of cash flow.

During Phase A, ROI is commonly based on top-level estimates of development and marketing costs, existing competitive pricing,

∎⋮ TABLE 4.3 *Application Domains for Product Development Paradigms*

Paradigm	Application Examples
Build, Test, Fix It	Home-built telescope, personal software application, applied research, one or few users, minimum need for documentation.
Staircase	Duplication of an existing product with minor changes and high inheritance: building of F-15 aircraft for export with avionics changes.
Staircase with Feedback	Requirements are clear and implementation technologies are bounded: exploration spacecraft, locomotives, instruments, automobile model upgrades, consumer electronics upgrades.
Early Prototype	Complete requirements are unknown but can be recognized by users: mature technologies, C^3I products, complex MIS products, new automobile models, novel mechanical systems, hydraulic systems, etc.
Spiral	Same as early prototype but tradeoffs between requirements and cost are unclear; greater risk with regard to implementation strategies.
Rapid Development	Complex, new products never attempted before or with unforeseen growth potential; requirements and design destined to evolve with operational experience: space station, novel command and control products, complex biomedical products.

and projected sales volume. Marketing costs include logistics costs. In Phase A planning, it is best to develop sales volume projections using a projection that ranges over the most pessimistic to most optimistic. For example, 2 to 7 percent of the market, or potential buying segment, in the first year, x and y percentages in the second year, and so forth.

DETAILED PLANNING GO/NO-GO REVIEW

The purpose of the Phase A review is to provide management with sufficient information to determine whether the product idea is viable enough to warrant further expenditure on construction of a detailed development plan.

The presentation of the review may run from half a day to a full day, depending on how much of the material outlined in this section needs to be covered. The main presentation should incorporate one to two slides for each item covered, possibly three as detail is required. Backup material in the form of additional slides for each item that cover the entire output of the team's efforts should also be available. That is, the main presentation should not be overly lengthy or cover detail that the management team may not need or want. The main presentation should be crisp, complete, and to the point. The backup slides are pulled out when elaboration is specifically requested, or management raises questions. Here is a guideline as to the content of the main presentation:

Topic	Number of Slides
Purpose	1
Content	1
Product Characteristics	
User need	1
Institutional fit	1
Partners	1
Sponsors	1
Constraints	1–2
Resources	1–2
Mission Statement	1
Design Concepts	1–3
Product Inheritance	1
Product Operations and Logistics	1–2
Development and Production Risk	1–3
The Development Paradigm	1
ROI Analysis	1–2
Recommendations	1

I know it is difficult to do a lot of work and not talk about it in great detail. This guideline, which comprises some 16 to 25 slides, is designed to succinctly tell the whole story in a timely manner without getting bogged down or getting off track. Rest assured, any need for detail will naturally emerge and a separate package of backup slides should be used to address further need for elaboration.

The presentation should be run by the Phase A team leader, who may call upon team members to offer remarks from the floor as needed. It is not a good idea in the main presentation to have too many team members come to the podium. If a subject is to be covered in more detail, the team leader has an opportunity to invite the appropriate team member to present the backup material and address questions.

This section on Phase A has provided a generic guideline for the basic elements that should be considered prior to further commitment for any product development. You may already routinely consider many, or all, of the points made here. But remember, things that are not thought of in advance have a way of coming up later. They surface as surprises or second thoughts. The goal is not to be surprised. Still, we are in phase A. Detail should be devoted to each element only to the extent required to gain confidence that product realization will satisfy a realistic customer need and that the concepts covered are sound enough to warrant the outlay of additional resources for detailed product planning, phase B.

ENDNOTES

1. N. B. Reilly, *Successful Systems Engineering for Engineers and Managers* (New York: Van Nostrand Reinhold, 1993).

2. See note 1.

3. B. W. Boehm, "A Spiral Model of Software Development and Enhancement," *IEEE Computer Journal*, May 1988; and B. W. Boehm and F. C. Belz, "Applying Process Programming to the Spiral Model," Proceedings of the Fourth Software Process Workshop, *IEEE*, May 1988.

4. W. H. Spuck, "Management of Rapid Development Projects," Jet Propulsion Laboratory, D-8415, April 8, 1991; and N. B. Reilly and W. H. Spuck, "The Products Engineering Process Under the Rapid Development Method," Proceedings of the Fourth Annual International Symposium of The National Council on Systems Engineering, San Jose, CA, August 10–12, 1994.

5. N. B. Reilly, *Quality: What Makes It Happen?* (New York: Van Nostrand Reinhold, 1994).

Phase B: Product Development Planning

5

═══════════════════

WHY PHASE B?

Dwight David Eisenhower was once asked by a reporter how he accomplished the amazing feat of marshaling the largest armada in history and successfully deploying it on European soil. His answer was memorable. He said, "Planning is everything." While our interests here are probably of a less bellicose nature, the point cannot be overstated.

Planning for any undertaking occurs only once. As soon as a plan is changed in an effort to improve it, then replanning is taking place. Plans for any complex undertaking are never perfect. A good plan, however, *maintains its structural integrity*. This is very important because, when surprises occur, accommodation, or modification, can occur without major perturbations in the structure of the plan. A good structure has a logical slot for anything unforeseen.

As was the case in Phase A, you may already be covering many of the issues to be discussed. In some cases, Phase B may be very simple and straightforward. But we will address the other end of the spectrum, where errors of omission are crucial. No matter what depth you go into in Phase B, remember: *It is the most important phase in any product development*. Planning is, in fact, everything. That's why Phase B. Now to the issues and this noble structure I have alluded to.

GENERIC PHASE B ISSUES

The generic issues to be addressed in Phase B are

1. Finalizing user needs
2. Building a work breakdown structure (WBS)

3. Organizing for success
4. Planning for control policies, which include

    setting product development priorities

    developing review and reporting structures and formats

    planning for technical margin management

    planning for documentation and formats

    planning for risk control

    configuration management

5. The logistics plan
6. The resource plan (workforce and costing)
7. The Product Development Team Management Plan
8. A detailed return on investment (ROI) plan

During Phase B, all of these issues are addressed in detail. The level of detail may vary, as appropriate, but the specifics need to be sufficient to enable management, upon review of the team's work, to make a sound decision as to whether to fund actual implementation of product development. This can be a major commitment. It is still possible after the more detailed exploration of Phase B to drop the whole idea. Before discussing these issues more completely, a few words on Phase B team membership.

Who's On the Team?

Again, team membership is determined by the what the team has to do. First we decide what issues are going to be addressed and then we pick the personnel with the right disciplines. Follow the lead of what you want to do. Here is a candidate list:

1. Whenever possible, include the end-user. If your product is for a single organization, have one or more users from that organization on your team. For general consumer products, have someone (probably from marketing) on the team who will faithfully act as the customer advocate.

2. Identify the person who will lead the concurrent product development team in Phase C at this time. The title may be Product Development Team Leader, or Chief Engineer, or Systems Engineer, Head Technical Honcho, or whatever, but we are talking here of an individual with broad-based systems engineering capability, a wide range of technical knowledge, excellent human relations skills, and a keen understanding that quality comes from motivated, empowered team members. He

or she should be on the Phase B team. That leader should take the first pass at a WBS, the desired organizational structure, and the drafting of the Product Development Team Management Plan, drawing upon the tenets covered in this part. When developed, these items should be presented to management for iteration and approval. But, their excessive iteration should not be required if the principles covered below under these topics are followed!

3. Someone from management should be on the team in order to develop the set of product development priorities after reading at least this part of this book.

4. There should be someone who understands logistic support for both development and operations.

5. You'll need a team member who understands bottoms-up integrated costing and scheduling and can effectively negotiate and iterate these with institutional division, section, and area management.

6. Expertise on risk identification and analysis for both development and production will be needed.

7. An experienced financial marketing type to develop a detailed ROI analysis is essential.

The candidate list entails fewer than ten good people. You may get away with less, depending on product complexity and the number of generic items you choose to cover. It is also possible that one individual may fill two or more of these roles. But you don't need or want many more. Now, let's look at the generic issues in more detail.

FINALIZING USER NEEDS

A good knowledge of the potential user and user environment *must* be a part of project planning. The finalization of user needs may easily involve corroboration of the Phase A concept through secondary and primary research, as required, to identify and document the complete set of user needs.

User needs are not functional requirements. Functional requirements are a set of highly structured statements of exactly what the system implementor proposes to do. Actual development of functional requirements comes later—at the outset of the preliminary design phase of implementation (Phase C). Functional requirements represent an actual signed off commitment and are generally a subset of, and derived from, a typically larger set of user needs.

The modern emphasis on quality has developed a useful tool to assist in this derivation. Quality function deployment (QFD) is a structured approach designed to translate the "voice of the customer" into high-level system requirements. The technique can also be used later in the system's engineering process to decompose higher-level requirements into lower-level specifications. A knowledge of the principles of QFD is useful during the gathering of user needs, particularly if these techniques are to be employed later. The first formal use of QFD techniques would occur after project start-up, during the early stages of preliminary design, when system functional requirements are developed.

The gathering of user needs at this stage may be less formal, but no less important. User needs are compiled from data gathered from existing documentation, market research, clinics, personal interviews, and so on. This is basically a complete wish list derived from the stated needs of the entire spectrum of the target user community.

Should there be any questions as to the data-gathering capability of the project planning group, then outside support should be enlisted to the extent required to acquire a complete understanding of the users' world. Technical people, and engineers in particular, have a marvelous penchant for developing systems that have little or no correlation to the true needs of the target user community. Failure to understand user requirements is extremely common and perhaps the most common cause of product failures. Products are built for customers—not for developers.

Examples are not hard to come by. Some years ago, an information system was installed in a hospital in England. The system concept was to gather and provide data at the nursing stations in each wing via CRT terminals and hard-copy devices. In the process of placing orders for supplies, clinical tests, and other services, data were to be integrated into a central computer for purposes of central billing, monitoring inventory, providing updated patient records, alerting admissions and other services of impending events, scheduling surgical suites, and compiling consumer demographics and other statistical reports. The CRT displays were menu driven.

One problem was the extremely limited space at the nursing stations. The engineers decided to put the terminals on carts with casters so that they could easily be moved about when access to specific areas was needed. Using casters was one of the only things they did right. After two weeks of training following "installation," the nurses had managed to push every terminal on each wing of each floor down to the end of the corridors and leave them there out of the way. After all that time and money, what in the world happened?

Throughout the entire preliminary user needs assessment and subsequent requirements generation phases of the project, not one nurse (frontline user) was involved in the product design. The menu-driven system required seven steps through medical ordering sub-menus just to order a particular medication. While menu-driven systems have many excellent applications, this was not one of them. Engineers never seriously questioned the applicability of this widely accepted technology to the nursing setting. Routine nursing clerical tasks were suddenly perceived as taking an excessive amount of time, with no apparent benefits. Benefits of the system design were oriented toward more efficient management. The motivation of most nurses is oriented toward quality and timeliness in direct patient care—not necessarily toward management. Fortunately, the previous vacuum-driven tube system through which documents were transported in cartridges was left in place as an emergency backup system. In short order, "the user" decided it was once again to be the primary system.

It is, of course, incredibly easy to pontificate about what happened in this example. User needs and requirements cannot be developed in an engineering vacuum. Unfortunately, what happened in this instance is not an isolated event. It is, in fact, a very common event. The whole purpose of the system was viewed as capturing data, and the engineers had a great way to do it. The leap to implementation of existing technology took precedence over any efforts to truly understand the total user environment.

For these kind of products, the term "user" actually refers to everybody in the customer organization. This includes top management, because it has to perceive a benefit and foot the bill. It includes middle management, because it has specific department- or division-oriented goals. It includes line management and supervisors because they have well-defined group goals. But most importantly, it includes the people who actually touch the system on a day-to-day basis because they typically have their own consumer interface goals. These goals can be, and are, extremely diverse. If a system design fails to meet all of them, it is clearly deficient. If the day-to-day user rejects it, it is a total failure.

The most prevalent reason for system implementation failure is that the product as delivered simply does not do what the customer perceived was needed. There can be many reasons for this—changes in requirements (sometimes called galloping requirements), drifting off course during design, and misinterpretation of what users say. But the most unforgivable reason is that the needs of ALL of the "users," including the day-to-day users, are not understood.

Detailed strategies for determining user needs are greatly influenced by the size of the user community. For systems that involve high levels of production, such as in the automobile and consumer electronics industries, it is not possible to interact with all of the users. User needs must often be gathered by statistical means through market research. One of the best ways to do this is to conduct interviews regarding existing products in an effort to determine likes, dislikes, and absent desirable features. It is also useful to field-test prototype concepts at trade shows, conventions, or clinics, when possible, to allow the customer to touch and feel.

Technical project planners are rarely expert in conducting market research and typically need to rely heavily on these resources within their organizations. Unfortunately, there can be considerable pressure brought upon the planners by other institutional interests. It is not uncommon for management and engineering personnel to be overly enamored with their pet ideas that are inconsistent with the findings of market research. If there is a choice, the customers' orientation should win out. If there is no choice, your organization has a quality problem.

When the customer community is relatively contained, the process may be simpler but not necessarily less difficult. Examples of such systems are an airline baggage system, a law-enforcement communications system, a government information or hardware system, or a system being developed for internal institutional use such as an oil rig or a corporate information system. In systems of this nature, the customer community is less diverse and can typically be directly approached as a whole. For these kinds of systems, the first step is to review all available documentation including any customer-generated requirements documents, specifications, internal memos, meeting minutes, engineering studies, or any other formal or informal material that refers in any way to the history of the perceived need. Usually by the time a user organization and an implementing organization first come together, a number of documents have been generated by the user in the normal development of their thinking to that point.

The Role of Requests for Proposals

Commonly, the first view of the customers' need comes in the form of a request for proposal (RFP). The RFP may or may not ask for development of requirements. If it does, you're in the user needs-gathering business up front. Often the RFP attempts to specify requirements. This can be an ambitious task for private sector and government organizations that do not routinely procure the type of system in question.

For example, a stock exchange that procures a modern transaction handling or data display system is likely not to have initiated such a procurement in a number of years, if ever. Members of the exchange cannot be expected to have top-notch technical RFP generators on staff and might be well advised to use a consultant.

Alternatively, branches of the military have had considerable experience in generating sophisticated RFPs due to their frequency of procurement. The same is true for organizations that routinely use suppliers. The experience of the procurer is usually evidenced by the quality and completeness of their RFP.

If the requirements seem complete, are measurable and distinct, and the level of experience of the requestor satisfies you, the need for a user needs-gathering phase may be greatly reduced or even eliminated. Still, we must not respond to RFPs that we know to be deficient simply to obtain business. If we feel user needs should be revisited or a more precise functional requirement should be generated in the interest of the customer, we should propose this course of action. We are not in business simply to obtain more business. We are in business to satisfy customers. It is imperative, therefore, that all parties clearly understand what a proposed system is to do.

The technical portion of a well-written RFP should contain a background statement, a succinct statement of the problem, top-level functional flow diagrams, system constraints, measurable requirements in terms of what has to be done, and a clear statement of work. If an RFP is received that lacks these attributes, it may be wise to propose that a thorough review of requirements through a user needs assessment be conducted as a first task.

Interviews and Questionnaires

Needs assessment involves interviews. Interview design and execution are specialties with which technical planners are generally not experienced. There are entire disciplines for this. A few general comments will give a flavor of the complexity involved. It is best to ask general, open-ended questions up front and gradually move toward specifics. As with all of us, respondents don't like to contradict themselves. Establishing concrete positions too early can limit freedom in later responses. Different statements of the same question are often inserted out of context in an effort to corroborate earlier responses. Open-ended sessions in which questions tend to migrate in their manner of presentation, or even change as the interviewer gets "on to something," must be conducted with great care. The sole use of open-ended questions does not typically result in a consistent database on which meaningful statistics can be applied. There is a definite art to the business. The best thing to do is get help with your questionnaire design from elsewhere in your organization or from an outside professional in market research.

An integral part of questionnaire design includes the design of the analysis of responses. The analysis design includes identification of the fundamental questions to be answered and the choice of statistics to be used to assure the statistical validity of results.

When you think you have a good questionnaire, review it with your customer. You will most likely have one questionnaire for top management and others for middle management, line management, and the hands-on product users.

Next, field-test the questionnaires with a representative sample of respondents. A limited, preliminary field test of questionnaires invariably uncovers deficiencies such as omission of specific questions related to unforeseen user perspectives, possible interpretation of questions in a manner other than anticipated, or the unintentional skewing of answers by the set of immediately preceding questions. These are very common deficiencies in questionnaires.

Following the field testing of questionnaires, the planned-for analysis should be carried out to further check that the responses will in fact provide the information you set out to get. The purpose of this exercise is to verify that the mechanics of the desired analysis are realistic, even though the statistical base in terms of numbers of respondents has not yet been established. The construction and execution of questionnaires is a typical area in which project planners may want to seriously consider the support of outside experts.

Determining True Needs

Table 5.1 offers an informative list of factors that contribute to the difficulty of gathering user needs and formulating requirements.[1] This list draws on the work of Barry Boehm as well as my own experience.

While one needs to be aware of all these impediments, item 5 is of particular interest. It is common for people to verbalize their problems in terms of perceived solutions. It is very important to be sensitive to this distinction early in the process. For example, the following statements are solution statements, not problem statements:

> "We need to put in an East-West teleconferencing network."

> "Management needs the ability to conduct quick-look, 'what-if' studies related to resource allocations."

██ **TABLE 5.1** *Factors That Make Determination of User Needs Difficult*

1. It's not always easy to define the customer or user.
2. Requirements are not always stated clearly or completely.
3. What the customers say they want may not be what they need.
4. You may be defining what you think they need, instead of what they want.
5. Problems are often verbalized in terms of solutions.
6. Implicit expectations may exist that are unreasonable.
7. You must talk to all segments of the user population.
8. If the users are not part of the planning and requirement analysis process, they are not likely to accept the product.

> "We need a computerized outpatient management system."
>
> "We need a dispatching system for our trucks."

Additional probing in response to such perceived statements of need is required to extract more generic statements of what the problems really are. Truer problem statements associated with the above examples might be:

> "Our East-West offices are not communicating in a timely manner."
>
> "Our resources are becoming oversubscribed. We need to increase our efficiency with regard to resource allocation."
>
> "We are not able to track our outpatients with sufficient precision to provide needed checkups in a timely manner."
>
> "We need to improve the efficiency of our use of vehicles in the field."

Authentic needs statements are identifiable because they logically give rise to further questions that result in needs that are measurable. When beginning with a true needs statement, one can sense this happening. Given the East-West communication problem, for example, one may now begin to pinpoint exactly what kind of information is not being exchanged in a timely manner, what its precise content is, what statistical response times are desired, and so on.

Avoiding "jumping to solutions" is not always easy. Ideas (solutions) related to our individual technological expertise naturally flourish in our minds. There are times when "solutions" seem so clear that it requires great discipline to slow down and concentrate on what the problem really is. At this stage of the game we are not a bag of solutions looking for problems. We are taking the first step toward problem definition. We are listeners who guide the customer toward generic problem understanding.

Here's a great example. A dear friend of mine once developed a very clever gimbaled platform with rapid mechanical feedback to maintain the platform at any desired orientation when subject to vibration. It worked great. You could mount a camera on the platform and shake it back and forth, and the camera image would not move. If you put a movie camera on the platform, and put the platform on the back of a bouncing truck, the image of anything following you down the road would stay rock solid in the frame. It worked great on helicopters, too. My friend immediately quit his job and went into business. He sold some platforms, but was out of business in two years. He spent all that time with a solution looking for a problem. He also had no real plan. If he had been user oriented, not solution oriented, he could have saved himself a lot of time and money. Not enough people needed his solution at that time. It's still a great idea, and it may have its day yet, but for my friend, the timing was wrong.

To further lessen the difficulties of defining the problem, the team should embrace the customers' cause by making an honest effort to visualize themselves in the users' occupational specialty. People will seldom support a design they are not involved in. To get them involved, we must get involved. We must learn customers' terminology and seriously prepare to be conversant in it. Then our questions and insights are meaningful and supportive of their problem environment. We must concentrate on *what* they need and postpone our urges to implant our pet technologies early in the process. It is important, of course, to guide a respondent should the user ask for features that are technologically infeasible, or clearly involve excessive cost. But our main duty at this point is to *listen*. Most user communities have one important characteristic that implementors don't. They know their user environment intimately.

Observation can often be as important as listening. This involves the "fly-on-the-wall" technique, in which users are observed using existing products in a effort to understand what they really do. The goal is to try to close the gap between what people say they do and what they actually do. Observation can also be a useful tool in coming up with new product ideas. Although these ideas are not new, more and more social scientists are getting involved in their application. They call themselves ethnographers and are out there observing everyone from teenagers, to factory floor workers, to retail sales people, to commercial fishermen—you name it.

The following points summarize the approach to understanding user needs:

1. Read everything you can find in preparation. Become as conversant as you can with the complete user environment. Ask questions. Change your shoes. Get smart.

2. Construct interviews. Respondents may include top management, middle management, and supervisors, but always hands-on users. If you have any doubt about your ability to construct a questionnaire, get help.

3. Review your questionnaire with someone with the appropriate experience.

4. For each basic function in the user environment, try to document top-level functional statements and needs for performance by measurable criteria, system design constraints, availabilities, and logistics support and information transfer in terms of data types, volumes, formats, response times, and tolerable error rates, and so on.

5. Field-test your questionnaire. Test out the analysis techniques designed to answer your questions. Fix what needs to be fixed. Try it again. Don't execute the questionnaire on a large scale until it works.

6. Don't be omnipotent. In their working environment, customers are smarter than you are.

7. Care about the people whose problem you are trying to understand. The more you care, the more they will like you. The more they like you, the better they will communicate with you. The more they communicate with you, the better chance of success. In this stage of events, you're the one who needs help.

8. Don't tell stories designed to impress the customer with your knowledge. The customer is the one with the knowledge. *Listen, Listen, Listen.*

User Constraints

System and product designs are often constrained by existing systems with which they need to interface. Rarely is a brand-new product built from scratch that is not impacted by some other system. This discernment of user constraints should be an integral part of gathering user needs and ultimate requirements formulation. User constraints are documented in the functional requirements document early in the implementation phase, should that phase be approved. Constraints are the only "how's" permitted at the functional requirements level. Typically, these constraints include interfaces to other existing product or systems to which the new product must conform.

One familiar user constraint is the need to interface with existing systems. Consider a hypothetical system to transmit a newspaper published in New York on a daily basis for printing and distribution in Chicago. A typical need may be to send newspaper copy after 1:00 A.M. upon completion in New York and to accomplish all copy transfer by 3:00 A.M., in time to run presses in Chicago. (This need will eventually set data rates.) Additional needs may include items such as acceptable error rates, system availability, data types, personnel, training, and the like.

Note that these needs state "what" must be done. The determination of "how" (satellite or land communications, use of aircraft, or carrier pigeons, etc.) is yet to be determined during options analysis and preliminary design. A typical constraint, however, in this example may be that the data interface in Chicago must use a particular computer model and the final product must be in a particular format on a particular storage media.

There are always cost and schedule constraints on any project. In some cases, however, they are absolute drivers. Examples are design-to-cost projects and space-borne systems that may need to be completed in time for specific launch windows of opportunity. Severe constraints such as power, weight, size, levels of performance, or design inheritance may also be imposed on commodities. While issues of this type may blur into classification as functional requirements, an early understanding of their potential for determining system feasibility should be sought. It is up to the planning team, through interaction with respondents, to determine if the constraint under discussion is truly to be a design limitation or driver. If it is, it should be identified and understood as a valid part of the user needs exercise and described at the level of detail required.

The User Needs Document

The user needs document provides a record of the user needs gathering exercise. A suggested outline is provided in Table 5.2. The introduction states the project name and includes designation of the planning team by organizational title. The purpose clearly states the objectives of the

▯: TABLE 5.2	*User Needs Document Outline*
1.0	Introduction
2.0	Purpose
3.0	Approach
4.0	Secondary Research
5.0	Primary Research
	5.1 Respondent Descriptions
	5.2 Analysis Design
6.0	Results/Conclusions
7.0	Sample Questionnaire(s)
Appendix A	Responses
Appendix B	Background Material

user needs study. The purpose should be related to the previously developed mission statement. Section 3.0 briefly relates the approach used, such as the use of secondary information sources, the use of questionnaires, and a top-level statement of the statistical tools used. Section 4.0 of the outline identifies specific secondary research sources. Section 5.0 covers the primary research activities through description of the respondent(s) (upper management, line management, hands-on users, and so on) and of the analysis design. The analysis design states the statistical methods used, and sample sizes, and ties the analysis to the purpose. Section 6.0 presents the results of analysis using charts and/or tables with supporting explanatory text. The section ends with a succinct statement of conclusions in bullet format. A sample of the questionnaire(s) is included as Section 7.0, and appendices are provided with questionnaire responses and any additional required secondary background material, such as articles and/or references.

The user needs document should be a stand-alone product that can be used for its intended purpose in support of project planning as well as for future reference by other projects.

THE WORK BREAKDOWN STRUCTURE

Now for the structure. Having defined and published a user needs document, the product development planning process next comes to the point at which a first cut at determining the details of what has to be done is made. The tool to be used is the work breakdown structure (WBS).

Now before you say, "We already do that and it works fine and I think I'll skip this part," hold on. I have seen a zillion WBSs as a worker bee and as a consultant, and there always is something that interests me. For example, when I am asked to attend reviews, I look for consistency. More often than not the items listed on the left side of a Gantt chart use different words than are on the WBS. In other words, something else is being scheduled. Then it turns out that cost accounting is being done on something else again, and performance measurement on something else, and metrics are measuring something else, and configuration management is being done on something else, and the organization is structured to do still something else, and everybody has all these different visions and versions of what's going on. I raise these points and I wreck the review and they don't ask me back. What is this?

Sorry, but a good WBS is an incredibly important tool. The right WBS, used correctly, will wield a profound influence on system architecture, internal system definitions, interface definitions, organization for implementation, product definitions, what gets scheduled and costed, requirements definition and flowdown, trade-off analyses techniques, testing, and the definition of disciplines required to succeed.

Unfortunately, there is no widely accepted standard for defining what must be done to accomplish product development. In fact, it is rare for any two people faced with this task to proceed in the same fashion. Some start out by trying to draw block diagrams. Others begin by making schedules, or drawing organization charts, or attempting to construct functional flow diagrams, or making drawings of conceptual designs. I have noticed that where one starts is often determined by one's professional background. Unfortunately, approaches such as these can be dangerous because they are not oriented toward the total picture. Typically these limited early approaches are oriented around the product alone. In doing this, the analyst automatically fails to engage in a thought process that gives appropriate attention to issues such as testing, logistics support, and management and systems engineering in a consistent top-down integrated fashion. An early concentration on the mission product alone leads to uneven emphasis with regard to the total system. It does not inherently provide a generic methodology to help one think of everything that needs to be considered and to decompose the work at consistent levels of description. Omissions, when and if discovered, run the risk of becoming ungainly appendages.

Further, these approaches rely heavily on one's experience. Experience, of course, is fundamental—but it can represent a serious drawback. In the absence of a methodology for a structured decomposition of work to be done at consistent levels, there is an overwhelming tendency to emphasize those technical areas with which one is most familiar, and to slight—or even forget—those areas with which one has less familiarity. This often results in excessive detail in one area and shallow treatment, or omission, in another area. Another good reason for the team approach!

So how do we do this? The decomposition process presented here is derived from hard-won experience. The work decomposition process is functionally the same as the processes encountered in decomposing a software system into computer software configuration items, programs, and procedures. It is also the same as the decomposition process carried out in knowledge engineering, or the decomposition process of structuring the organization of a hierarchical database. It's all the same stuff.

What we seek is a top-down process in which each successive level of detail consists of independent parts. Each level of detail is directly traceable upwards and downwards to an entity at a higher level or lower level of decomposition. Everything has a specific, well-defined place. It can be a difficult process requiring many iterations, but when done correctly the structure will exhibit a peculiar beauty. That beauty is most severely tested when an unforeseen addition is required long after the structure is in use. The robustness is evident when the new addition does not perturb the structure but comes immediately to a resting place that is logical and consistent—a resting place that is unmistakably identified by the original structure itself.

It is always best that the first cut at a decomposition process be carried out by a single person, in isolation, with sufficient time and with a minimum of pressure. It is expedient for the concurrent product development manager (team leader) to make this first cut. There will be time for review and critique by the project planning team.

The WBS presented in this section is a generic one. As the term *generic* suggests, the structure applies to the development of the entire class of "systems"—that is, any product or system. The basic structured thought process involves addressing each item in the generic WBS at successive levels of detail. The generic guide is intended to generate complete horizontal consideration of the entire system at each level of detail *before proceeding to the next level of detail*. Review and iterations between levels are, of course, likely to occur. At any given point in time, however, the principal concentration of effort is devoted to completion of a specific level.

▯⦂ **TABLE 5.3** *A Generic Product Hierarchy Nomenclature*

Level	Nomenclature	
1	System / Product	
2	Segments / Subsystems	
3	Elements	
	hardware	*software*
4	Subsystems	Programs
5	Components	Program Components
6	Subassemblies	Modules
7	Parts	Procedures

While the material presented here is primarily devoted to product development, it is also of interest to recognize that, at the top level, WBSs for production, assembly systems, operations, or for system retirement are comprised of basically the same generic ingredients. System retirement considerations can also profoundly affect the way a system is designed. It is understood, however, that products are designed for operations and that operational considerations also drive development requirements for the mission product and its operational logistics support.

The work is to be broken down at each level of decomposition. It doesn't matter what each level is called. One convention for naming each level is shown in Table 5.3. You may use your own, but you must be consistent. For example, in many organizations the top level—the system or product level—is referred to simply by the product name, for example, the Taurus. The next level down refers to systems, for example, the power train system. If you are in the space business, it is useful to talk in terms of a flight system and a ground system (typically called segments), with each having systems, then subsystems. Any scheme is acceptable as long as the terms for each level are consistently used at the appropriate level only.

Enough talk. Let's get to it.

Remember Phase B is planning for Phase C. There are only five things you need to do in Phase C product development. Someone has to do the programmatic stuff—that's *management*. Someone has to make sure everything comes together technically, on time and in budget, that is, do *systems engineering*. Someone has to provide parts, spares, training, transportation, facilities, and maintenance—that's *logistics support*. Someone has to be sure everything works—that's *test-*

ing. Someone has to actually *build the mission product. Everything you need to do falls into these categories.*

After a few overview comments, we will address each of these in more detail.

Management covers programmatic issues at each level. At level 1, or the product level, project management products consist of such items as the Project Management Plan, the Configuration Management Plan, control policies, and so on. Similar management plans are products for each lower level—segment, element, subsystem (or whatever).

The concurrent product development team has the responsibility for the other four items. Systems engineering is responsible for the technical success of the fielded mission product and its support mechanisms, or it's interface to support mechanisms. There is a systems engineering role to be fulfilled at each of the levels of product decomposition. For example, the System Functional Requirements Document is a product at the systems level, as are system test requirements. Subordinate requirements, detailed designs, specifications, test plans and procedures, and so on are typically produced at the lower levels.

Logistics support addresses how the development of the mission product is to be supported and how the mission product is to be supported in the operational environment. The latter influences design of the mission product. Again, these support issues arise at each level of system decomposition.

Testing refers to the functions of validation and verification of product performance and the performance of its subsystems and elements, and so on. Testing includes parts testing, unit testing, integration testing, system level testing and, for one-of-a-kind systems, acceptance testing. Pre-ship and post-ship testing may also be called for.

The mission product is simply the complete entity that is to be created. The mission product is identified at each successive level. This decomposition is referred to as the mission product breakdown structure (MPBS). The MPBS consists of the hardware, software, auxiliary equipment, and integration equipment items that make up the system product itself.

Many of the products included in the WBS are documents. This is particularly true for management and systems engineering, although documents are produced in all five major WBS categories at all levels. Further, in each category, material covered at one level may be sufficient to cover lower levels as well. Needless repetition should be avoided in all documentation to alleviate the complexity of change control. In this scheme, lower-level documentation should be produced to cover issues unique to that level only and to make use of references to higher level documents, whenever possible, to avoid repetition. Each of the generic products discussed below, however, should be considered for each level in the interest of completeness.

The following subsections discuss each of these five basic components of the WBS in more detail.

Management

The lines between systems engineering, systems engineering management, project management, and product development team leadership can be subtle, and their delineations are often a matter of style and experience, or organizational policy. The style is often set by upper management and sometimes by project management. It is of extreme importance that the specific roles of project managers, deputy project managers, product development team leaders, software managers, and managers at each level (segment, element, subsystem, whatever) be well defined and synergistic with regard to the management WBS items. We will see that a good WBS clearly defines everyone's role and responsibilities.

Management activities include

- preproject planning support
- product development planning support
- setting of top-level budgets and schedules
- setting and/or approving product development priorities
- providing top-level goals to all three product teams and empowering them to do their work
- conducting reviews and making go/no-go decisions

Systems Engineering

Systems engineering is the focal point for the organization and operation of the concurrent product development team. This includes technical responsibility for the development of preliminary and detailed requirements, requirements flow-down, designs, fabrication, system test requirements, and transfer to operations.

Generic systems engineering activities consist of

- providing management support
- organization and leadership of the concurrent product development team

Management support includes support in the construction of any of the management activities listed above. The second bullet encompasses a great deal. The concurrent product development team is the focal point of all systems engineering activities. Technical team activities include development of requirements, designs, identification and resolution of system issues, trade-off studies, selection of design tools and aids, interface definition and control, margin management, development of test requirements, specialty and concurrent engineering activities, and the making and monitoring of day-to-day operating schedules for complete implementation.

Generic systems engineering products that appear at appropriate levels of the systems engineering WBS are

- the Product Development Team Management Plan
- functional requirements
- design requirements or specifications
- software requirements
- software designs or specifications
- interface documentation
- test requirements
- configuration management support documents
- reports and reviews

The term *system* is often reserved to apply to a total system under development. However, subsystems, elements, and so on are also systems in themselves. Systems engineering must, of course, take place at each level where development is taking place. Systems engineering during development typically stops at the point at which procurement of components, subassemblies, off-the-shelf software, or parts takes place. For example, a radio, an accelerometer, and a chip are all self-contained internal systems. But they may not require systems engineering if they are to be inherited or procured for a larger system under development. At those levels where systems engineering is required, the above listing provides generic systems engineering products that should be considered at each level of the WBS.

Logistics Support

Logistics support refers to those WBS items required to support the development and operations of any system. There are two distinct logistics support efforts that must be considered. One is the development support system, and another is the operational support system.

The former is generated to support the Phase C implementation process, and the latter is a part of the design process for support of the mission product itself once delivered for use. Generically, these are the same, but their content is quite different. Operations logistics support items are a part of the mission product breakdown structure, since these items are an integral part of the mission product.

Generic logistics support activities consist of the planning for, design and fabrication of, or the acquisition of

- supply support
- test equipment
- transportation and handling
- technical data packages
- support facilities
- personnel and training
- maintenance

These components are coordinated through the concurrent product development team to assure that all design impacts of logistic support are taken into account.

Logistics support products include

- logistics support plans
- all hardware and software items for each of the logistics support components during development
- reports and reviews

Testing

Testing takes place at all levels, from the mission product level to the parts level. At the product level, there may be three types of testing. These are subsystem integration testing, complete system level testing, and user acceptance testing. System level testing may involve both pre-ship and post-ship testing prior to user acceptance testing. Testing at each level below the system level (segment, element, and so on) consists of integration testing of the next lower-level units and internal system testing at that level prior to integration testing at the next highest level.

Testing activities include

- development of parts test plans and procedures
- development of subsystem test plans and procedures at each level where required

- development of integration test plans and procedures at each level where required
- development of user acceptance test plans and procedures at the product level
- execution of integration testing at all levels

Generic testing products that appear at appropriate WBS levels of the testing portion of the WBS are

- parts test plans and procedures
- integration test plans and procedures
- system level test plans and procedures
- acceptance test plans and procedures
- test reports and reviews

Note that the test requirements are a product of systems engineering at each level.

Mission Product Breakdown Structure

The mission product breakdown structure (MPBS) follows the product decomposition into logically smaller units with a *consistent* degree of detail at each level. The segment level is often used to denote significant physical system partitions. Examples are a ground segment versus a flight segment, or a remote segment versus a base station segment. In systems where such delineations are not appropriate, as in decomposing an automobile, or a refrigerator, it is common to eliminate the segment level in the decomposition process. Some prefer to go directly from the system, or product, level to the subsystem level. Any of these modifications are acceptable in tailoring a given product to the process. The breakdown, however, must use consistent, agreed-upon terminology and exhibit consistent degrees of detail at each level.

Table 5.4 presents one example of the hierarchical thought process associated with the first three levels of the MPBS.

At this level in the thought process, the analyst concentrates on basic *functions* needed to meet the stated user needs. The level of consideration is important. We are at the top levels of a top-down process. At these levels, basic technologies and disciplines are considered. If we perceive that light detectors are required, we do not put down light detectors—we put down sensors. If we perceive that radios are required, we do not put down radios—we put down communications. Similarly we put down navigation, if required, at this level—not inertial navigation platform. Nor do we put down items such as diesel engines or ram-jets, but instead we put down propulsion, or power plant. In short,

▯ TABLE 5.4   *A Generic Mission Product Breakdown Example*

System	Segment	Element
System		
	Segment A	
		Sensors
		Communications
		Navigation
		Power Plant
		Chassis
		Consol
		————
		Assembly Equipment
		Auxiliary Equipment
	Segment B	
		————
		————
		————
		Assembly
		Auxiliary Equipment

at this level we are still addressing fundamental functionality. The "how" comes later. The last two items listed in Table 5.4, assembly and auxiliary equipment, warrant particular consideration. Assembly equipment is that equipment employed to integrate, or connect together, the other level-three functional items. This includes such items as cables, conduits, racks, mounts, and the like. Every system has assembly equipment. Assembly equipment will *always* be an item at every level below level 1 of the WBS. Engineers who design and build functional internal systems at various levels are typically not interested in how they get connected with other internal systems to form a larger system. The builder of the WBS and, of course, the product development team must be.

Auxiliary equipment is that equipment to be used by more than one internal system or by personnel in the operational environment. This includes such items needed for environmental control, production and distribution of electricity, common storage space, common work space, security, shelter(s) and other commonly used facilities. It is rare that a system does not have some type of auxiliary equipment.

▐ TABLE 5.5	Mission Product Breakdown Structure for a Locomotive	
System	**Element**	**Subsystem**
Locomotive		
	Power	
		Engine
		Intake
		Exhaust
		Lubrication
		Starting
		Cooling
		Fuel
	Traction	
		_____
		_____
		_____
	Control	
		_____
		_____
	Structure	
		____
		____
	Assembly Equipment	
	Auxiliary Equipment	

If none occurs to you at this stage in the thought process, you will still do well to include it for completeness at each level. It is easier to remove items from a structure than to insert them after the structure is formulated.

While there is flexibility in the execution of the process, there must still be significant logical rigor. For example, consider the product breakdown structure for a locomotive. Table 5.5 presents a product breakdown in which the terms _element_ and _subsystem_ are used at the second and third levels. Note the consistency of description at each level. The breakdown asserts that a locomotive consists of four product elements, plus assembly and auxiliary equipment.

Alternatively, consider a product breakdown structure for a product (system) to provide cellular space-based data and voice communications for earth-based users. Let's call it SpaceCom. Table 5.6 employs the system, segment, element, and so on, terminology. Again, note the

TABLE 5.6 *Mission Product Breakdown Structure for a Spacecraft*

System	Segment	Element	Subsystem	Component
SpaceCom	Flight	Spacecraft	Payload	Communications
				Power
				Housing
				___
			Structure	___
			Attitude Control	___
			___	
			___	
		Launch Vehicle		
		___		
	Ground	Assembly		
		Auxiliary Equipment		
		___		
		Assembly Equipment		
		Auxiliary Equipment		

consistency of the product breakdown at each level. Note something very interesting here. The whole mission of SpaceCom is to provide communications. But it doesn't even show up until level 5.

As with the breakdown of systems engineering products, the mission product breakdown typically stops at the point at which procurement of components, subassemblies, off-the-shelf software, and/or parts take place.

The complete WBS is built by assembling all products for management, systems engineering, the mission product, logistics support, and integration and test at each level. Figure 5.1 provides a generic template for this process applicable to each successive level. The management, systems engineering, logistics support, and integration and test of a set of products at a given level takes place at that same level. The management, systems engineering, logistics support, and integration and test of a specific product within the collection of products identified at level N takes place at the next level down. This is not as difficult as it sounds. In the locomotive example, a management function takes place at the power element level as well as the next level down, for example, the engine subsystem level. Similarly, systems engineering, logistics support, and integration and test functions also take place at each level.

An example of this complete decomposition should be helpful. All of us are involved in different products and systems. But one common product most of us are familiar with is the automobile. Let's take for an example the Phoenix automobile discussed in the Phase A section. First, we break down the mission product as shown in Table 5.7. In the example, the engine has been broken down into parts. A similar level breakdown for a hydraulic steering system might include parts such as a reservoir, hoses, fluid, and a structure. Note, however, that the generic breakdown does not yet specify whether the steering system is to be hydraulic, mechanical, electrical, or what. Note also the inclusion of auxiliary equipment and assembly equipment.

With this initial work accomplished on the mission product breakdown structure, we can now build complete work breakdown structures at each level for the automobile. Figures 5.2 through 5.7 provide a number of examples of the breakdown. Note the numbering scheme used in the figures. The numbers 1 through 4 are permanently associated with management, systems engineering, logistics support, and integration and test, respectively. Levels of the product breakdown are designated by letters (A, B, C, etc.). In the example, power train is designated by A and the chassis by the letter B. The designations are carried consistently to each successive lower level. For example, in

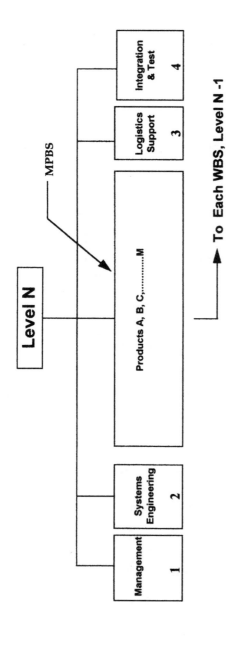

Level N

MPBS

Management	Systems Engineering	Products A, B, C,............M	Logistics Support	Integration & Test
1	2		3	4

→ **To Each WBS, Level N -1**

Note: The management, systems engineering, logistics support and integration & test of a collection of products at level N takes place at level N

The management, systems engineering, logistics support and integration & test of a specific product identified at level N (A, B, C, - - -) takes place at the next level down

The construction of lower levels stops when engineering stops and procurement of components, subassemblies or parts starts

Figure 5.1. The Generic Product Development WBS.

▪▯ TABLE 5.7   *Automobile Product Breakdown Structure*

Product	System	Component	Part
Automobile			
	Power Train		
		Engine	
			Cylinder block
			Cylinder head
			Valve train
			Pistons
			Connecting rods
			Crank shaft
			Manifolds
		Transmission	
		Starter	
		Distributer	
		Fuel injection	
		Engine controls	
	Chassis		
		Steering	
		Brakes	
		Tires/Wheels	
		Suspension	
		Exhaust	
		Fuel	
	Body		
		Structure	
		Interior	
		Exterior	
	Electrical		
		Power	
		Alternator	
		Power distribution	
		Diagnostics	
	Auxiliary Equip.		
		Heating	
		Ventilation	
		Air conditioning	
		Entertainment	
	Assembly		
		Body	
		Paint	
		General assembly	

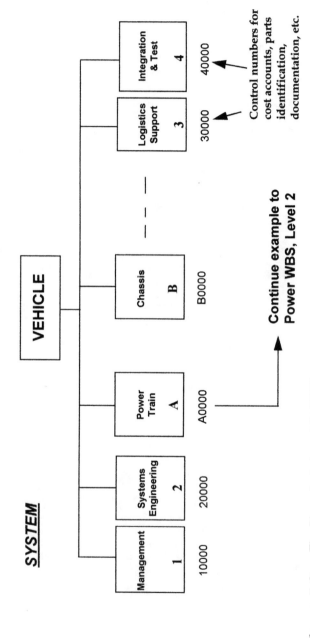

Figure 5.2. The Phoenix Automobile WBS.

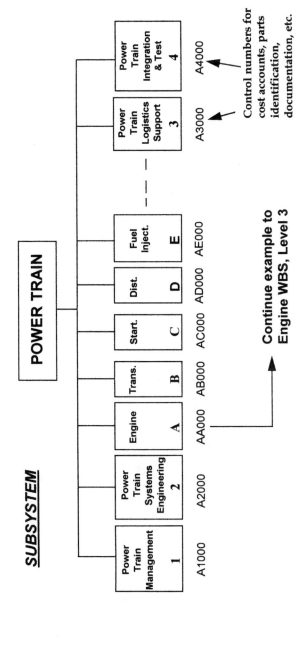

Figure 5.3. The Phoenix Automobile WBS (Con't).

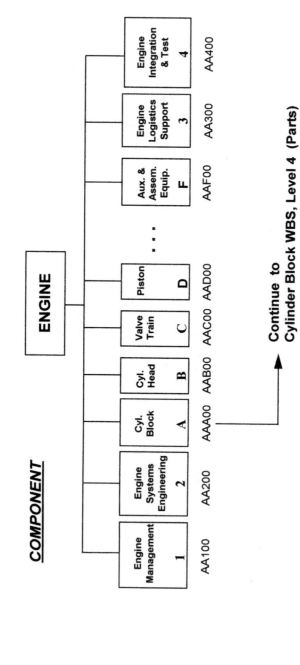

Figure 5.4. The Phoenix Automobile WBS (Con't).

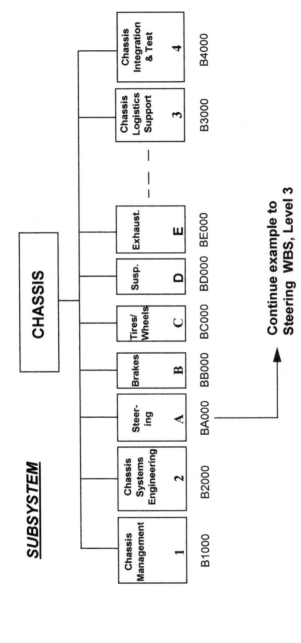

Figure 5.5. The Phoenix Automobile WBS (Con't).

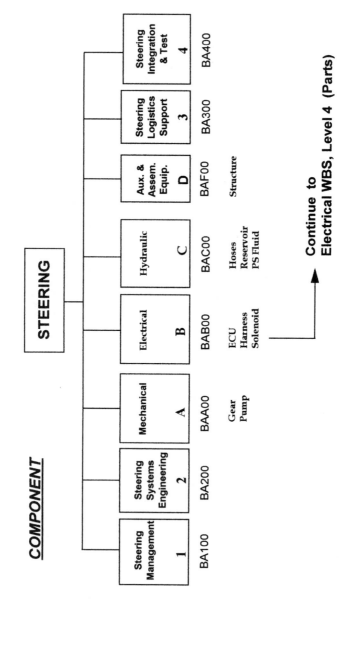

Figure 5.6. The Phoenix Automobile WBS (Con't).

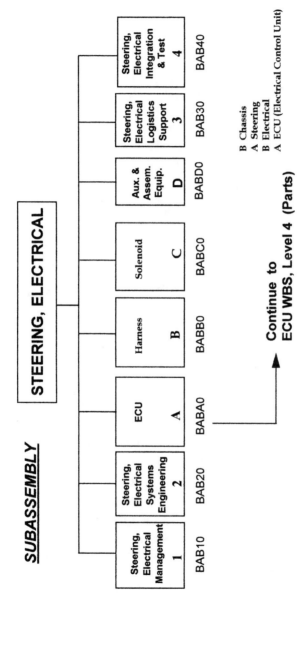

Figure 5.7. The Phoenix Automobile WBS (Con't).

Figure 5.4, the sequence AAAOO represents system A (power train), component A (engine), and part A (cylinder block). Similarly, the sequence AA200 represents systems engineering for the engine. The scheme can be adapted so as to provide a consistent and easily readable set of designations for account numbers and all documentation, including drawings and parts lists.

Don't panic. We are not creating a zillion jobs. We are not staffing here. We are trying to address "what" generically has to be done. Multiple assignment of items to a single individual is possible and common. Filling slots comes later.

This exercise should clarify the importance of initiating system definition with construction of the MPBS and a complete WBS at each level. Functional block diagrams do not provide a complete view of the entire system. Similarly, a complete understanding of what needs to be scheduled can be gained only by viewing the total system as represented by the total WBS. Finally, the WBS must be a living document that is constantly updated and improved as the design and implementation process matures. Structuring of a WBS is commonly prone to two types of errors. The first is compromising the top-down nature of the effort by mixing different levels of detail. The second consists of errors of omission. The generic WBS guide is designed to minimize the probability of committing either type of error at all levels.

The purpose of the generic WBS is to assist in consideration of everything that has to be done to accomplish system definition, design, implementation, and test. The generic WBS provides a guide to where each activity and product should be assigned. The structure presented is a sound one that has evolved from and survived many applications. There is, of course, a certain flexibility in how it is used; however, it should not be necessary to deviate very far, if at all.

In deciding where a particular line item belongs in the WBS, there are five simple level two questions to ask:

1. Is the item in question a piece of the prime mission product?
2. Is the item in question to be used for purposes of validating, verifying, or testing any part of the mission product?
3. Is the item in question to be used to support development of the mission product or to support operation of the delivered mission product?
4. Is the item in question related to the definition of project management in your environment?
5. Is the item in question related to the definition of systems engineering in your environment?

In this scheme of the WBS, everything has a place. If there is any confusion as to where something belongs, identify it with one of these five categories first. Consider the following real-life examples.

EXAMPLE 1: RADIO TOWERS—Suppose we have a system where radio towers are to be used to support communications repeater elements. Building towers and building radios are two quite different disciplines. While radio and repeater engineers will certainly supply the radio sets themselves, they typically don't build towers, nor may they even be concerned with power or transmission cables that need to go up or down the towers. Are the required towers part of the operational support system, or an element of auxiliary equipment under the mission product, or are they a part of the communications subsystem of the mission product?

Since the radio system link design typically calls for radios and repeaters to operate at specific heights, the associated towers are an integral part of the communications mission product. The towers are not a part of the assembly equipment line item because the towers are not to be used by more than one subsystem. The towers are not a part of logistics support because they are an integral part of the end product item and are directly required for the mission product to meet its functional and design requirements. Thus, under communications on the MPBS would appear the further breakdown of communications elements to include radios, repeaters, cables, tower mounts, and towers as a minimum. The appropriate disciplines to meet these level four needs would then be defined and sought as parts of the communications subsystem team.

Alternatively, if the towers were to support not only communications elements, but say navigation elements as well, then we may assign towers to the assembly equipment line item at level three since they would support more than one subsystem. Because we only want one team designing towers, that team would be scheduled in one place under auxiliary equipment and would be assigned the task of meeting requirements of both subsystems.

EXAMPLE 2: CABLES—Cables can seem to get complicated very quickly. Here are some approaches.

Cables between computers and peripherals, for example, clearly are a self-contained part of the computer and are a responsibility of the data processing subsystem. Cables servicing or interconnecting different level three items should be a part of assembly equipment. Such cables include power distribution from mobile power units in fixed/mobile systems and cables devoted to information flow between subsystems. In complex systems where there are many cables, it is expedient to devote considerable effort to cable design and layout philosophy. For example, in systems calling for a confusing myriad of cabling, a rational design approach is to standardize pin connections and cable lengths into, say, short-, medium-, and long-length cables. This can be done for both power and data cables. In this scheme, there are only three (or six) kinds of cables in inventory, which can greatly relieve the replacement problem encountered when a large number of different special cables would otherwise be required. Each cable contains an identical superset of wires required by all subsystems. The female connectors at each subsystem chassis employ only that subset required by that subsystem. These design issues would properly be assigned as a level four item under the level three assembly equipment line item for the mission product. In complex systems, such as tanks for instance (ever been in a tank?), cabling may warrant elevation to a subsystem in itself. In any case, interfaces should become very clear. The cabling subsystem interface ends at the male ends of each cable and the communications, or navigation, or whatever, interfaces lie at the female receptacles on their respective unit housings.

EXAMPLE 3: AIR CONDITIONING—Air conditioning is an integral part of environmental control. It's position in your WBS, again, depends on thoughtful review of its use. Air conditioning for an operational command and control center could be assigned as part of auxiliary equipment supporting more than one subsystem of the mission product. In an automobile, it is also under auxiliary equipment. Alternatively, air conditioning for a development laboratory supporting design could be considered as a part of the facilities line item under development logistics. The important point to get here is that the generic WBS presented has a logical place for everything and can even accommodate individual conventions and practices. Also, notice how it helps you think of stuff you might not have thought about.

While it is evident from these examples that there is some flexibility in the categorization of specific line items, it is important to bear in mind that your hand is not completely free. Your choices are limited by the generic structure itself. Should any particular item fail to fall neatly into place, then this is a very positive signal that extra-thoughtful consideration is required. Slow down and think your way through. Walk away from it and come back later, as necessary. If you are not sure either exactly where each item belongs or the rationale for its placement, then the people who will actually be doing the work will never know.

It is worth taking a moment to emphasize the knowledge gained through the simple diligence of building a sound WBS. Having devoted sufficient time to its thoughtful and thorough construction, you will have gained considerable insight in nine specific areas:

1. You can now sketch out top-down system block diagrams. Top-down means you begin by drawing a single block, representing the entire mission product, with top-level annotation of inputs and outputs. The next level of block diagraming exhibits a single block for each mission product line item at level two. Then the next one at level three, and so on. The terms you use for inputs and outputs must be consistent and complete. For example, at one level you might have as inputs, "User Inputs." At the next level, user inputs would be shown in more detail, such as, "Inquiries, Updates, Administrative." This process is the same as the process of building a data dictionary.

2. As you develop the system block diagram, the internal systems (subsystems, elements, etc.) should become evident.

3. As internal systems begin to crystallize, so do the required interfaces between them. If it is obvious at this point that there are overriding considerations for configuring internal systems and top-level interfaces differently, then you may do so. Recognize, however, that any changes made at this point will invariably change your WBS. Iteration is, of course, quite acceptable and even expected—but consistency must be maintained.

4. The WBS, by it's nature, suggests how you must organize to accomplish your technical goals.

5. The WBS tells you directly what you need to schedule at all levels.

6. The WBS identifies those elements on which you choose to do cost accounting.

7. The WBS helps identify high-risk areas for which effective mitigation strategies need to be developed.

8. The WBS identifies the disciplines required to accomplish your goals—the same disciplines required for representation on the concurrent product development team led by the product development team leader.

9. The WBS identifies items that you may wish to place under configuration management.

Invariably, it is easier to gain insight into any of the above nine areas by first structuring the WBS. The reason is simple. Our primary interests in the product/system definition process are completeness and the maintenance of consistency.

The generic WBS is the recommended tool for the complete definition of what must be done. It is considerably more difficult to initiate such definition by, say, building a block diagram of something without a structured concept of the complete task at hand. Four-fifths of what must be done (management, logistics, testing, and systems engineering) are not covered adequately by mission product–oriented block diagrams.

Similarly, it is difficult to schedule something until one knows what to schedule. If the first step in organization is to build schedules, then clearly two things are being attempted at once. That is, one is trying to identify items to be scheduled and scheduling at the same time. Scheduling is a part of resource planning, not WBS construction. It is a lot easier, and a lot less dangerous, to do one thing at a time.

It is not uncommon to observe practicing engineers and managers initiate the process of product definition somewhere in the middle and then expand in both directions. In the absence of any widely accepted standard, the prevalent attitude (excuse) is, "Everybody is a little bit different, and that's the way I do business." This is nonsense and can be dangerous. The purpose of the generic WBS is to suggest a standard by which a logical structured approach can be executed.

Organizing for Success

Organizing for success requires some understanding of how organizational structure will impact your planning and staffing, and subsequently your ability to succeed. Let's review two classical organizational structures. Then, we will discuss four actual case studies in which organizational structure caused product development failure. Once we've seen real examples of how not to organize and why, we will discuss how you *should* organize.

There are two generic types of organizational structures, the functional organization and the matrix organization.

The functional organization is organized for relatively stable mission products such as the automobile, steel, or home entertainment industries. The organization is structured around each product and the functions needed to achieve product realization. The functions needed are repeated for each product. Figure 5.8 depicts this basic structure and gives an example of functions that might be needed. The particular functions are determined by the mission of the organization—whether in government or the private sector.

FUNCTIONAL ORGANIZATION - PRODUCTS TYPICALLY STABLE

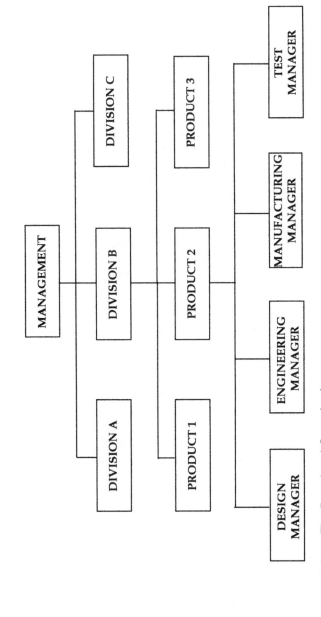

Figure 5.8. The Functional Organization.

113

The matrix organization is organized for major mission products that vary over time, or when products are routinely tailored to each customer, such as research or high-tech enterprises. Its structure is based on a pool of separate technical capabilities. As Figure 5.9 shows, the pool is "tapped" as product developments come and go. Project managers decide on who is needed and call upon the technical resources to provide the appropriate personnel from across the matrix.

These generic types of organizations do not represent rigid categories. In fact, mixes of the two can be found. In reality, the implementation of these generic structures involves a great diversity of expression—some good, some not so good. But they are the generally accepted generic types.

The functional organization tends to have more stability with regard to both project structure and personnel turnover from project to project. It also fosters the creation of project-oriented turf (my sandbox, your sandbox). The matrix concept exhibits less conformity with regard to project structures because project managers tend to have more freedom in the way they organize in adapting to projects of different natures (not always a good idea). Also, it is generally easier for personnel to move from project to project in a matrix organization. However, the matrix organization fosters the creation of discipline-oriented turf (your sandbox, my sandbox).

In either case, organization has a profound effect upon the ability of programmatic managers and technical managers to maintain visibility. Although seldom admitted, organizational structures alone are frequently among the major reasons for system development failures. Surprisingly, it is common to find product development groups that are not organized to do what they intend to do.

The following four case studies describe actual project organizations, each of which was involved in a costly product development failure. The failures are directly attributable to the inability to perform management and systems engineering roles due to organizational flaws alone. The examples are taken from both functional and matrix organizations. A single critical flaw across all examples is then identified and a project structure that is designed to avoid this problem is suggested.

Case 1

The organizational structure is this case (Figure 5.10) was devised by the project manager. A first observation of the arrangement is that it is not clear who is in charge. Systems engineering, software management, environment & QC, and resource management all appear as

MATRIX ORGANIZATION - PRODUCTS CHANGE THROUGH TIME

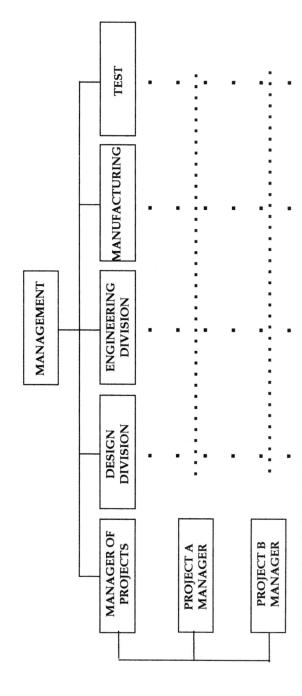

Figure 5.9. The Matrix Organization.

115

WHAT'S WRONG WITH THIS ORGANIZATION ?

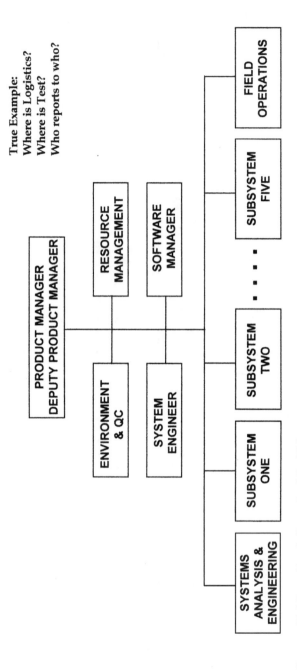

True Example:
Where is Logistics?
Where is Test?
Who reports to who?

Figure 5.10. Don't Organize Like This.

"stem winders" to project management. In other words, they all report to a single line. In this design, the path to authority for element and subsystem cognizant engineers (COGEs) is equally ill-defined. In practice, when a COGE identified an issue perceived to have system impacts, he or she first decided whether it was primarily a programmatic issue, a software issue, or a systems engineering issue. The COGE would then approach what was considered to be the appropriate person. Unilateral decisions were often made that effected all functions without the timely inclusion of counterparts and without concurrent product development team coordination. The concept of the concurrent product development team as focal point for the handling of all issues was slowly eroded.

Further, the chart confuses the relationship between the systems engineer and the software manager. The systems engineer has technical responsibility for the complete system, which includes both hardware and software. This indefinite structure eventually led to confusion as to who had technical authority. The confusion, in turn, gave rise to discussions among the workers who were trying to determine who had what responsibilities—an issue never clearly addressed by project management. The deputy project manager, in an effort to allay growing confusion, began to take on more technical authority, which resulted in further confusion of roles. The deputy manager's perception was that he was strapped with a pack of power-hungry animals. Organizational deficiencies were never considered.

Note also that the structure does not conform to the generic work breakdown structure for the following reasons:

1. Environment and QC are two different things. While it may be reasonable to include QC as a staff function, environmental issues should be handled through specialty engineering representation on the concurrent product development team, as these matters directly effect the system design.

2. There is no clear indication of how testing functions are to be accomplished. There is no organizational testing function.

3. The closest function to operations logistics support is the box entitled field operations. There is no clear provision for the remaining logistics items of personnel & training, technical data packages, transportation & handling, facilities, or support equipment to sustain those operations.

The roles of stem winders were substantially hindered on this project to the point that it was unilaterally canceled by the sponsor when it became evident that a cohesive product design was not emerging on schedule. Management found fault in "an inexperienced staff" and an unreasonable customer. The word *introspection* was apparently not in their lexicon.

Case 2

Product development was originated for a software-intensive system to automate data gathering and reporting at the national level. Figure 5.11 graphically summarizes the condition. The project initiator was in a separate department from the three departments in which the resources actually existed, Departments A, B, and C. The problem was viewed as a software problem, and a contractor was selected and brought in. The contractor did exactly what the project manager asked: developed a software system.

The project manager knew "exactly" what data inputs and outputs were needed to solve *his* problem. It was assumed that when the system was developed, Departments A, B, and C would want to use it. Why not? The manager had a great idea. But there was no design team formed—hence no real user interaction.

The system was completed and installed, but never used. Why? Simple. The project manger did not establish a product planning team, much less a product development team, to organize the effort around a user-oriented focus, which included all the system users. It turned out that the required data which presumably existed in Departments A, B, and C were not organized in the form required by the unilateral software design. Further, the department heads of A, B, and C, having not been involved, simply found ways not to provide the data. Their response was, "What is this, more work?"

In short, the system was never used because from day one the project manager did not understand what the "system" was. Hence, he could not organize properly to solve the problem. Any product development effort must include constant and intensive user interactions. There are people out there! There was no WBS. None of this bothered the vendor, who should have made some attempt to avoid proceeding without a better understanding of the user community. The whole effort was doomed from the start.

Case 3

Case 3 adhered to the classic staircase systems engineering approach. The paradigm was followed faithfully with the development of requirements, specifications, the holding of all proper reviews, and so forth. All the right steps were executed, but by the wrong people.

ORGANIZE TO INCLUDE THE USERS

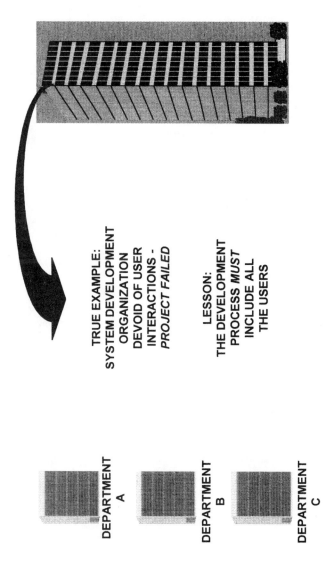

TRUE EXAMPLE:
SYSTEM DEVELOPMENT
ORGANIZATION
DEVOID OF USER
INTERACTIONS -
PROJECT FAILED

LESSON:
THE DEVELOPMENT
PROCESS *MUST*
INCLUDE ALL
THE USERS

DEPARTMENT
A

DEPARTMENT
B

DEPARTMENT
C

Figure 5.11. Don't Organize Like This (Con't).

119

Because the project was so "important," the communication between the two organizations was handled solely by upper management. Top management knew the generic process but was unable to empower responsibility. It was their pet project and who knew better than they exactly what had to happen?

The real users were at the bottom of the user's organization, and the real implementors at the bottom of the contractor organization. Users and implementors communicated with each other through a number of levels of management in each organization. The communication path is depicted in Figure 5.12. There was no nitty-gritty weekly concurrent product development team structure through which the actual users and developers could interact. The implementors and users of the system were not empowered to communicate, much less work together. They didn't even know each other's names.

After ten years (count 'em—ten years) of effort, a zillion meetings on the top floor, and an expenditure of thirty million dollars, the system did not work properly upon delivery. Today, they are still "working" on it.

Case 4

Case 4 involves a very large matrix organization with discipline-oriented divisions. The major roles of each matrix division are fairly well defined. However, they are dispersed geographically so that each of the divisions has found it necessary, over time, to implement overlapping capabilities at their own sites. Similar functions appeared everywhere, such as systems engineering, testing, and so on.

Top management was also geographically isolated, but what is more significant, exercised programmatic control only over the matrix divisions. Technical authority over day-to-day events was distributed— delegated to each center of expertise.

New programs were assigned by top management to a matrix division, or in the case of big projects, to a group of divisions. Geographically isolated divisions could also call upon each other for support. The success rate was higher when the assignment was made to a single division, or when support was agreeably solicited among divisions. In fact, the divisions worked fairly well by themselves.

When a large project was assigned to multiple divisions, turf issues typically arose. This is because the members exhibited a common characteristic of matrix participation: The members thought all divisions were equal, and some of them thought they were better. When such an organizational structure assigns the role of, say, systems

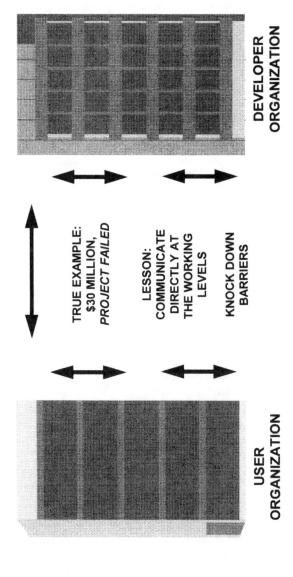

Figure 5.12.   Don't Organize Like This (Con't).

121

engineering, to one of the divisions, the other divisions reacted with comments such as, "They're not going to tell us how to do systems engineering; we have our own people to do that." The result was that no one was really in charge. There was no central day-to-day technical component with the authority to manage all the divisions. The structure simply does not provide the authority for technical management roles and integrated planning to take place under these conditions. The responsibility exists—but not the authority. There is a difference between empowerment and anarchy.

The result was that big multimember projects were placed in serious jeopardy from day one solely due to the fact that they were not organized to succeed. The organization I allude to still operates that way today.

Organizational Flaws

These four cases may appear extreme. Unfortunately, they are not only real but involve some of our nation's most prestigious private sector and government organizations. What is more impressive is that, in each case, the product development and organization managers found reasons for their development failures that were totally unrelated to their organizational structures and management styles. The inward look is, evidently, extremely difficult. These structures, and ones like them, are created over and over again and continue to be sustained, even by managers that confidently aver "I would never make a mistake like that."

Note also that in each of these cases there are common threads. None of the organizational structures were staffed to respond to a sound WBS. All of them fostered significant turf barriers. And none had a structure to help them consider all the issues involved.

The organizational issue has come under increased attention with the advent of concurrent engineering and modern quality concepts. In the conventional nonconcurrent approach to development, each organizational entity tends to determine requirements in a relatively isolated environment. Each has its own turf. Thus, the various functions of developing user needs, designs, manufacturing needs, assembly systems, testing, marketing strategies, operational support, and so on have often lacked vital coordination. Each entity would basically pass results of their work "over the wall" to the next. In addition, it has not been unusual for suppliers to be sought out after the design process. The integrated product development concepts, using corporate-wide expertise in a team setting, call for total coordination of all of these functions independent of the development paradigm in use. Complete consideration of all factors that influence product development also includes design integration with suppliers.

Organizations initially undertaking concurrent engineering often begin by organizing design and manufacturing functions under one new department. This has probably been the case because design and manufacturing have traditionally been areas where considerable contention has been visible within the physical organization. Designs that have not adequately considered producibility issues are often bounced back and forth incessantly until time runs out and management simply orders a "go ahead."

While reorganization is a beginning, it addresses but a part of the overall problem. All aspects of user needs, design, engineering, manufacturing, assembly, testing requirements, marketing strategies, operational support, close supplier involvement, and the like should be considered. Organizationally, the focal point for concurrent engineering should be the product development team, under the leadership of the product development team leader.

The lesson for programmatic and technical managers is simple. Don't take assignments in organizations that are not patterned to accommodate a structured team approach around a sound WBS. You must be empowered to run your show. In short,

DON'T TAKE RESPONSIBILITY WITHOUT AUTHORITY

If you are not given sufficient authority in a technical or management role, you will not be able to effectively implement or control the processes and/or the tools and personnel required to sufficiently maintain programmatic and technical completeness and visibility.

Recommended Organization

We have reviewed some actual examples of organizational structures that significantly contributed to product development failures. Figure 5.13 outlines a project structure that is more amenable to sound programmatic and technical management principles for Phase C product development. Although the model can be adapted to fit particular product development needs, the farther you stray from this model the greater the risk of problems.

The structure shown in Figure 5.13 has the following very positive characteristics:

1. The concurrent product development team (PDT) function is clearly in line to project management. While project management is ultimately responsible for everything, the PM

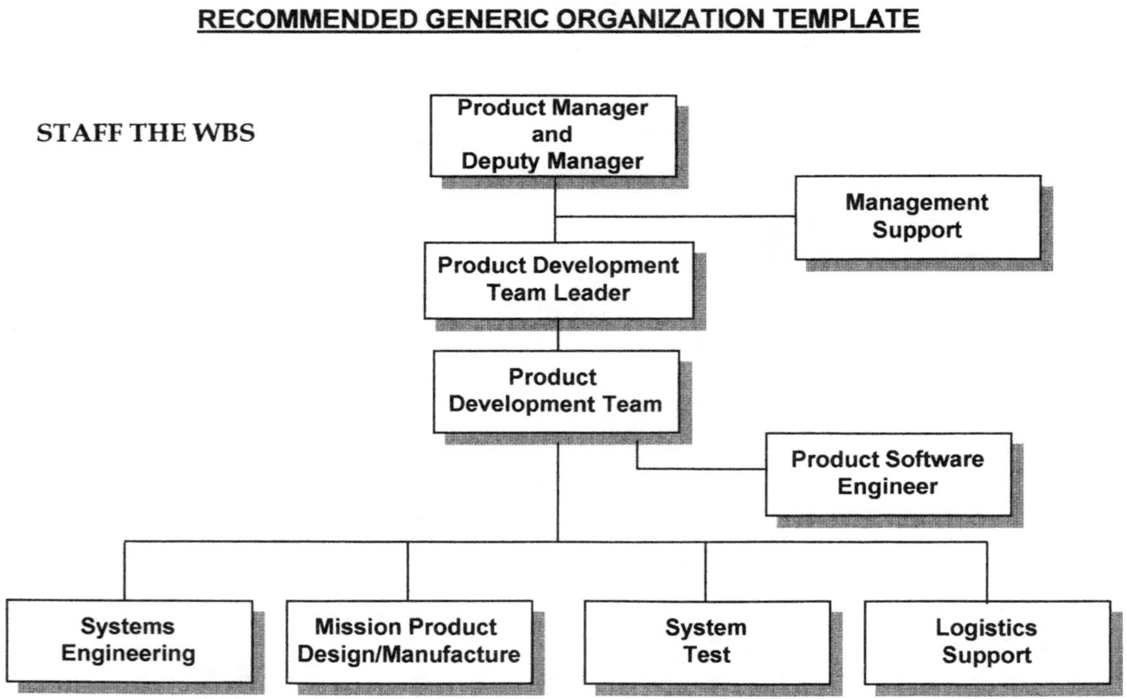

RECOMMENDED GENERIC ORGANIZATION TEMPLATE

STAFF THE WBS

Figure 5.13. A Good Generic Organization.

office is primarily concerned with programmatic and upper management interface issues on a day-to-day basis. Complete technical responsibility lies with the concurrent product development team function. The authority and empowerment in the structure is clear.

2. Software systems engineering clearly reports to the concurrent product development team. The lead software engineer is also a member of the team. The product development team is superior to all of the four major functions at lower levels. This is because total system software engineering not only involves operational software for the end mission product but is also commonly integrated with software issues related to logistics support such as operational training and maintenance, as well as testing. The mission product software is often integrated with these other software needs. The software systems engineer, as a part of the product development team, must be given the authority to address the total system software picture—not just the required operational software. In this scheme, the systems engineering functions of the PDT retains the complete system overview including software systems engineering.

3. The concurrent product development team is clearly delineated and is superior to the four basic lower-level elements. The concurrent product development team is clearly organized and managed by the team leader. It is also the clear focal point for the execution of concurrent systems engineering.

4. The five basic generic WBS components of management, the mission product, logistics support, system testing, and the concurrent product development team staff are clearly represented and clearly structured.

5. The product development team leader is the sole entity responsible for reporting to management. There is not a lot of confusion here.

A slightly different—but still consistent—view of product development organization can be taken directly from the WBS model. Figure 5.14 expands our thinking to recognize that the generic WBS functions occur at every level of corporate activity. The first thing we see is that the five basic WBS elements are repeated at the corporate level, the division level and the model level. We know these functions repeat below the model level (subsystems, etc.), but now we will go in the other direction.

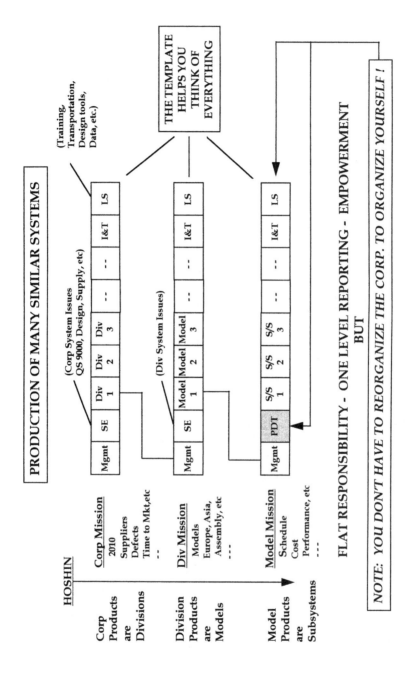

PRODUCTION OF MANY SIMILAR SYSTEMS

THE TEMPLATE HELPS YOU THINK OF EVERYTHING

(Training, Transportation, Design tools, Data, etc.)

HOSHIN

Corp Products are Divisions

Corp Mission
2010
Suppliers
Defects
Time to Mkt,etc
--

(Corp System Issues
QS 9000, Design, Supply, etc)

| Mgmt | SE | Div 1 | Div 2 | Div 3 | -- | I&T | LS |

Division Products are Models

Div Mission
Models
Europe, Asia,
Assembly, etc

(Div System Issues)

| Mgmt | SE | Model 1 | Model 2 | Model 3 | -- | I&T | LS |

Model Products are Subsystems

Model Mission
Schedule
Cost
Performance, etc
- - -

| Mgmt | PDT | S/S 1 | S/S 2 | S/S 3 | -- | I&T | LS |

FLAT RESPONSIBILITY - ONE LEVEL REPORTING - EMPOWERMENT BUT

NOTE: YOU DON'T HAVE TO REORGANIZE THE CORP. TO ORGANIZE YOURSELF !

Figure 5.14. A Corporate View For Product Development.

Consider the corporate level. We have corporate management, naturally. Sample issues at the corporate management level are: where are we going to be in the year 2010 (market shares, domestic or foreign markets, etc.)? or reduction of suppliers with closer ties, or defect reduction goals, or time-to-market goals, or consistency of control policies and plans, or streamlining acquisition, or mergers, or concentration on core capabilities, and so forth. Divisions can be considered as principal corporate products. Are there corporate system level issues across divisions? Yes. For example, QS 9000 implementation, or the use of integrated automated design tools across divisions, and the like. The testing function generically asks: Is what we are doing working? At the corporate level, test means: Are the management policies, division strategies, and logistical strategies working? Are our customers happy? This includes the whole bailiwick of market research.

Are there logistics issues at the corporate level? Yes. How about training across divisions (the corporate products), reporting formats, transportation efficiencies, just-in-time philosophies, and the like?

Now how about the division levels? Division products are specific models (such as the Pontiac Division's Firebird, or a television manufacturing division's 9-inch color, AC/DC model). Are there division management issues regarding models? Yes. If corporate says we want to expand our foreign markets, then division responds with where (Europe? Asia?), and figures out what modifications of existing models, or what new models, are required. The Chinese don't want or need American cars as we know them. Are there systems engineering issues across models? Yes. How about production, assembly, common subsystems, parts across models, and the like? Are there test and logistics issues across models? Yes.

A Way of Thinking

Figure 5.14 places the Product Development Team at the mission model level. *Here lives the specific product development team in the systems engineering role.*

What are we saying here? We are talking about a way of thinking in the interest of completeness. We are talking about generic levels of responsibility *and* authority. What should be done and who should do it? We are saying that a given operational level of an organization can be regarded as a system in itself. A system has a product that needs management. It also requires addressing of system issues across the product (*at that level*), logistics support, and a way to determine if the system is working properly.

Why is this so simple and so nice?

1. It helps you think of everything that has to be done at all levels.

2. It provides a template for where everything belongs.

3. Roles in the organization are very clear (people, including you and me, love to know exactly what they are supposed to do and an opportunity to do it).

4. Career paths within and across disciplines are clear. For example, future managers will want to move across the systems, product, test, and logistics functions. Alternatively, setting your goals on corporate logistics means moving up through product and division logistics levels. Good training, right? (No more kicking unqualified people upstairs.)

5. Reporting paths are clear (who and what they have to talk about).

6. Corporate sets corporate goals, divisions translate these into model decisions, and the model level develops specific products. This is called Hoshin planning, where each level basically sets goals and policies at its own level and stays out of the way of the next level down. This is also called empowerment. This is what breeds self-respect and excitement about coming to work and has an inherent positive effect on product quality.

Incidently, this approach blends well with all the talk about flattening organizational structures. There are three levels shown in the diagram (there are actually more as each mission product is broken down). But, there is a flatness with regard to responsibility because each level is responsible to and reports to only the next level up. It is solely the next level up that has complete responsibility for everything done.

Despite the fact that I love Figure 5.14 (it might be the best in this whole book), I must emphasize that I am not asking you or anyone else to reorganize. I am simply presenting a way of looking at organizational structure that is oriented around what has to be done to develop a product and subproducts at each level. You don't necessarily have to reinvent, or reengineer, or reorganize the whole corporation to effectively use a product development team. The product development team is at the model level. Just do it at the model level.

Reorganization, or reengineering, for its own sake is not always a good idea, although the maneuver is more time honored than we may think. Does the following sound familiar?

▄▄▄▄
We trained hard—but it seemed that every time we were beginning to
form into teams, we would reorganize. I was to learn later in life we
tend to meet any new situation by reorganizing, and a wonderful
method it can be for creating the illusion of progress while producing
confusion, inefficiency, and demoralization.

–Petronius, A.D. 66
▄▄▄▄

Reengineering is a modern term for a process that may or may
not result in reorganization. Reengineering is not directed at improvement of existing processes. That's process improvement, which is making things you are already doing better in a given environment.
Reengineering is something else. It involves forgetting everything you
are doing now, starting with a new piece of paper, deciding what it is
you want to do, and then organizing to do just that at all costs. This
may or may not call for drastic change. If no change is called for after
the exercise, so much the better, but you have still gone through the
reengineering process.

These thoughts on organizational planning are designed to present
a way of thinking to assist in covering all the bases. They simply suggest
that the way to organize is to staff the WBS—which is a complete template, a top-down statement, of what must be done. It is true that a
proper organization in itself does not guarantee success in product
development. But it is certainly true that when we are organized to do
something other than what we want to do, we are severely handicapped.

CONTROL POLICIES AND PLANS

Control policies and plans (CP&P) can be developed in parallel with
planning for Phase C organization. These policies and plans represent
the complete set of tools and strategies for the control of all aspects of
Phase C product development. Development of CP&P begins with the
identification of top-level product development priorities. Once established, these priorities are used to guide development and flow-down
of requirements and design trade-off decisions. They also provide a
clear understanding, in advance, of the order in which compromises
may need to be made throughout development. Once you have established top-level priorities for Phase C, all other CP&P components may
be developed in parallel. After a detailed discussion of how to develop
priorities, we will discuss these additional CP&P components that can
be developed in parallel. These additional components consist of:

- *Reviews*—These consist of formal and informal review strategies. Examples of formal reviews are planning reviews, product requirement reviews, preliminary design reviews, critical design reviews, test readiness reviews, and pre- and post-ship reviews. Less formal reviews consist of inspections, walkthroughs, and routine assessments of progress during working meetings.

- *Reporting*—Reports are presented or distributed in support of the evaluation function. Report examples are executive summaries, cost reports, schedule reports, and technical process reports. Reports cover both planned and unplanned events.

- *Margin Management*—This involves identification of crucial commodities, such as response time, power, weight, size, and so on, and the development of strategies to manage their allocations across the product design.

- *Documentation*—CP&Ps for documentation specify what documents are to be created, and the responsibilities for generation and management of documents. Documents are products and appear as items in the work breakdown structure.

- *Risk Control*—This term refers to the specification of plans and procedures for identification and control of risk during product development. Control of risk includes the development of preplanned backup strategies.

- *Configuration Management*—The plan for configuration management includes configuration identification, configuration control, configuration auditing, and status accounting. CM includes plans and procedures to facilitate management of unanticipated, nonscheduled events. Controls include, but are not limited to, engineering change requests, engineering change orders, waivers, liens, and action items.

Note that these CP&Ps are oriented toward product development. Issues dealing with such items as safety and risk control during system operations should surface as functional requirements that are generated by the concurrent product development team during Phase C implementation.

The following paragraphs discuss each of these PC&Ps in more detail.

Defining Product Development Priorities

Here's an exercise. Leave your office, or cubicle, for a few hours and walk around asking people involved in any product development what their priorities are. If you get a different answer from management, marketing, engineering, design, production, fabrication, assembly, procurement, testing, potential customers, finance, and all the rest of the folks involved, your ship could be tighter. Products are frequently conceived, designed, implemented, and delivered without a clear and consistent set of priorities in the minds of all parties involved. This is because no one has established them. It is important to devise and negotiate a set of priorities directly with the customer (or through effective market research). Management, of course, should concur so that the major factors that will influence the establishment of requirements and guide potentially crucial design decisions are clearly understood by all parties *in advance.* The setting of priorities is also of value in achieving a common understanding of how and why initial designs, or even requirements, may need to be altered should compromises need to be made.

Further, the establishment of clear priorities provides a framework for consistent behavior on the part of programmatic and technical management, the entire development team, the users, and all others associated with the project. These priorities also aid in communication with management. The establishment of consistency throughout the entire process is a principal goal of planning. The incorporation of agreed-upon priorities sets a tone for order at the highest level for all involved parties. Further, I can't begin to tell you how much time is saved in endless discussion when priorities are well understood up front.

Priorities are derived primarily from listening to the "customer." They are also balanced by an engineering knowledge of what is achievable at an acceptable level of risk. The "customer" includes the ultimate product users, system operators, maintenance personnel, supervisors, and all levels of user management—everyone who touches or is touched by the product. Typically, they all have different priorities.

An important question to ask the customer community at all levels during definition of user needs is "What do you want most out of this product?" or "What aspect of this product is most important to you?" Driving issues such as cost, schedule, or performance generally surface in response to such open-ended questions if they are truly foremost in the respondent's mind.

Should clear top-level patterns not emerge, further specific probing on issues such as cost, performance, schedule, availability, flexibility, maintainability, operability, transportability, safety, mobility, and so on should be initiated.

Recall that the mission statement includes these considerations. The content of the mission statement provides an ideal starting point for the determination of priorities. For example, consider the mission statement we developed earlier for a new-model automobile. It went like this:

▭▭▭▭
••••••••••••

Market research has identified a user need for an economical, medium-performance, comfortable, youthful, and sporty-appearing vehicle with easy access for the mature driver. No such vehicle exists in today's market. We will provide integrated concurrent engineering resources to include suppliers, dedicated internal cross-departmental assets, dealer service organizations, and continued user feedback for the timely introduction of a competitively priced automobile that is economical to operate and meets customers' expectations for appearance, comfort, and performance, with a time-to-market of two and a half years.

••••••••••••
▭▭▭▭

Note the last sentence of the statement, and in particular the words from "timely introduction" on. Timely introduction means schedule. Competitively priced and economical to operate means costs. These words are followed by characteristics of appearance, comfort, and performance. In this example, logical development priorities for the automobile are:

1. Schedule
2. Cost
   a. Price
   b. Operations
3. Appearance
4. Comfort
5. Performance

It is not necessary to pin down the exact limits of tolerance in advance on each item. The listing is meant to be more of an agreed-upon guide for design and design tradeoff activity. When any requirement is in jeopardy, it's impact must always be discussed with the customer, or with a team member representing the customer.

The concept of priorities is also of value in avoiding surprises and potential disappointment on the part of the user community. Consider, for example, a deep space probe system whose major purpose is to perform scientific exploration at a distant planet, such as the recent Mars Rover mission. The spacecraft will be designed to carry a number of instruments. But, surprisingly, the number of instruments to be carried—that is, the science-gathering capability of the platform itself—may be of the lowest priority. Consider that the target planet inherently imposes a specific launch window for a given launch vehicle, propulsion system, and mission trajectory design. If the launch window is missed, there will be no mission—hence no science. Further, if the size and weight of the probe exceed certain limits, the launch vehicle will be inadequate—hence no launch. If the communications element consisting of transmitters, receivers, and antennas cannot provide sufficient effective radiated power to reliably send the science data back, then the system is degraded, or fails.

Now I know that most of us are not deeply involved in the space science business, but the example is of interest. It is interesting because we have a situation in which meeting a launch date with stringent specifications on mass, size, and communications capability—to name only a few factors—takes precedence over the number of instruments on the spacecraft, even though the purpose the entire enterprise is to accomplish science. Thus, *science becomes one of the lower priorities,* and it is entirely possible that the number of science instruments may be reduced on the final spacecraft over the number originally intended in order that the mission be achieved at all.

Do some of you deal with the military? How about this real example? Again, it is included because the outcome may not be entirely intuitive. Some years ago there was a call for a Rapidly Deployable Mobile Gun System, or, RDMGS. The purpose of the system is to allow for rapid deployment anywhere in the world of a highly mobile weapon capable of attrition (military speak for destruction) of the most formidable of adversarial tanks. In this example, rapid deployability translates into a requirement to be shippable by any and all means of existing transportation, including the ability to be lifted by a helicopter. Lethality requirements for the specific threat to be engaged suggest that at least a 75 mm to 105 mm primary weapon be mounted on the RDMGS.

One hard reality of these requirements is that a small tank that could be lifted by a helicopter would not be capable of carrying a significant amount of weight devoted to protective armor. A main battle tank typically weighs some 50 tons, while an RDMGS-type vehicle could not weigh much more than 20 tons, total. Thus, the survivability of such a vehicle must clearly depend on its lethality, mobility, and crew training. In short, survivability is a function of the element of surprise, ability to acquire, identify, and attrit quickly and accurately and then move on, as opposed to its protective armament. It is further assumed in this example that discussions during the establishment of user needs indicated that the military senses a pressing need for such

❏⋮ TABLE 5.8	Priorities for the RDMGS
TRANSPORTABILITY	Ability of the RDMGS to meet deployment requirements by fixed-wing aircraft, helicopters, and road, rail, and amphibious means.
LETHALITY	Ability of the RDMGS to identify, engage, and defeat specified threat targets.
MOBILITY	Ability of the RDMGS to meet specified maneuver requirements on land and water.
IOC DATE	Ability for the RDMGS to meet it's initial operational capability date.
SUSTAINABILITY	Ability of the RDMGS to meet requirements for reliability, availability, and maintainability.
COST	Ability to field the RDMGS within a life-cycle cost limit.
SURVIVABILITY	Ability of the RDMGS to avoid and/or withstand threat.

a vehicle, thus the initial operational capability (IOC) date is firmly established. It is also clear that operational logistics support requirements for the RDMGS will be quite different from those associated with the more conventional use of a tank-type fighting vehicle. The RDMGS will be substantially separated from any tangible support mechanism during execution of a given mission, but each mission is also likely to be quite short. These factors have an impact on system availability design tradeoffs.

With these considerations in mind, the set of priorities proposed for the RDMGS is shown in Table 5.8. The following rationale is suggested for the RDMGS priorities.

Transportability is ranked as the highest priority, since the ability of the RDMGS to be rapidly deployed is fundamental to the mission concept. Succinctly—if it can't get there, it can't do anything.

Once the vehicle is successfully deployed, the major purpose becomes to defeat the threat in order to accomplish the mission. Thus, lethality is ranked second.

Given it's lethality, the field commander must be able to move his assets quickly to desired positions over specified obstacles. Mobility is deemed an important part of the hit and move concept, but it is subordinate to transportability and lethality. The ability to field this asset in a useful time frame is considered the next important priority. Because of the short duration of proposed missions, coupled with the hard fact that the RDMGS will be a vulnerable entity, vehicle sustainability is rated next. Cost, always an important factor, is rated as subordinate, within limits, to all of the higher priorities.

Due to the inherent light weight of the RDMGS and the current state of anti-armor weaponry, it is decided that excessive expenditures of money and design activities to achieve a high degree of survivability for the RDMGS would have a low probability of paying off. From the mission perspective, the burden of survivability shall be carried indirectly by the higher-priority design characteristics of lethality, mobility, and the ability of the RDMGS to be rapidly removed from the chosen arena of engagement upon termination of the mission. It is also clear that with existing technologies, no amount of money can purchase sufficient protection for a vehicle of the required weight in the required mission setting.

Clearly the RDMGS platform is to be designed for a dangerous and highly aggressive mission. After careful, stern and objective consideration, system design priorities have emerged that place the lives of the combatants themselves at the lowest of priorities—a condition that, mentioned out of context, would not likely to be acceptable. Military assets are not generally designed with a low priority on safety. In this case, the realization of priorities emphasizes that the survivability issue is clearly relegated to a need for a high degree of training, operational coordination, and the use of the element of surprise. I remember this situation because I was the one who realized and developed these priorities at the outset. The colonel I interfaced with found these very interesting, quite rational, and agreed that it was very important for everyone to understand these priorities up front in the development. The example is a realistic one and emphasizes the importance of thoroughly considering priorities early in the concurrent product development team process.

Detailed priorities for consumer products are quite different from the previous examples. Principles of the modern quality movement, for instance, dictate that the highest priority for consumer products is customer satisfaction and that among the lowest of priorities is development cost. This tenet is rapidly replacing older, management by objective concepts such as "decrease production costs by 10%" or "increase sales by 5%," which are internally directed priorities for product development. For consumer products, the term "customer satisfaction" typically translates into performance, which, in turn, maps onto the meeting of functional requirements including maintainability. In any case, it is not sufficient to simply state that "quality" is a number-one priority. Priorities, like requirements, must be clearly measurable. The goal should be to delineate a set of specific, well-defined, and measurable priorities related to such issues as performance, availability, reliability, delivery date, operability, cost, and so on. The concepts of quality function deployment are designed to address these issues in detail.

Priorities can never be taken for granted, and unless they are clearly stated will constantly differ among all participants throughout the design nd implementation process. It is of substantial importance to establish and agree upon a set of priorities in the early going, if for no other reason than to avoid unnecessary and extensive discussions and confusion as the implementation unfolds. As the examples have shown, it can also be a very enlightening process in causing one to consider the consequences of meeting top-level requirements in a structured manner that brings factors to the fore that would surely be less well-defined without going through the exercise.

Examples are useful, so let's take another one. Consider priorities for a videocassette recorder (VCR). Suppose that market research has discovered that most available VCRs are too complicated for a considerable segment of users to make full use of their capabilities. Folks are saying things such as, "My son has to come over to make it work," and "I'd pay up a little if I could run it—too many teeny buttons." We look at the mission statement that succinctly states where we are and where we want to go.

▬▬▬▬

Market research has determined a need for a VCR that is significantly easier to operate. Customers indicate that they will pay slightly more for this feature. Further, this condition has been common since the inception of these products. We will develop, test-market, and produce a new VCR model with intuitive ease of use as it's main feature. The new model will meet or exceed existing model reliabilities, serviceabilities, and performances.

▬▬▬▬

That's it. What is a good list of priorities that respond to this mission statement?

1. Ease of Use
2. Cost
3. Delivery Date
4. Product Availability
5. Performance

Consider this rationale. Ease of use is agreed to as number one. If the unit turns out to cost a little more, it's OK. While we don't wish to be too long in getting to market, ease of use and cost are still more important, because we perceive that the ease of use issue has existed for some time and is likely to continue to be a problem for competitive models. Product availability (A), in the engineering sense, is given by

$$A = \text{Mean time to failure} /$$
$$(\text{Mean time to failure} + \text{Mean time to replace})$$

This means we want a high mean time to failure and a low mean time to replace—that is, good reliability and a good service organization. In our example, existing models have an excellent record in these respects.

Finally, all other performance issues come last because our existing models are also quite good with respect to performance. This does not mean that availability and performance are not important. It just means that, in the development of this product, we will inherit existing availability and performance capabilities. We will concentrate our time and money on the first three priorities.

We also note that the product is to be test-marketed. This means our Phase B planning includes planning for a relatively polished prototype during Phase C, the development phase. These plans should include how the VCR is to be test-marketed—in homes of employees and their families, or with friends, or at trade shows, or at advertised clinics, or all of these.

The setting of priorities is very important. Here's why:

1. They are an important tool to be used in requirements flow down and in conducting trade-off analyses during the development phase.
2. They allow early identification of dear commodities such as cost, schedule, response times, availability, operability, power, weight, and the like.
3. They provide for top-down traceability.
4. They establish *in advance* agreements between management, developers, manufacturing, marketing, finance, customers, and so forth.
5. They save an inordinate amount of time and discussion throughout the development cycle. They are a major asset in getting to market faster.
6. They serve nicely when using QFD.

Quality Function Deployment

Quality function deployment (QFD) is a complementary method for determining how and where priorities are to be assigned in product development. QFD is designed to translate customer demands into each stage of product development. The intent is to employ objective procedures in increasing detail throughout development.

Figure 5.15 presents an example of the technique. The product is for a skip-loader. The left-hand column lists features desired by the customer, referred to as customer demand quality. The features are broken down into three levels. For example, the top level demand for cost is more specifically stated at level two as operational cost, and at level three as costs for service, fuel, trade-in, and so forth. The quality plan is then derived as shown on the right section of the figure. The first column of the plan rates importance of each feature to the customer.

The second column lists estimates of the current state of skip-loader products for our own company and for two competitors.

The third major column under the quality plan section reflects our plan for quality, the resulting improvement rate over what we are doing now, and a metric called the sales point. For example, we plan to have a service rating of 5, which is what the customer demands (wants). This is good because we, and our competitors, are rated at 4. The improvement rate ($d=c/b$) is 1.25. The sales point metric is 1.5 for a double circle entry, 1.2 for a single circle entry, and 1.0 if there is no entry.

The last major column of the quality plan section gives the weight and demand weight of the feature. For the service feature, the weight is given by the product of the importance, improve rate, and sales point, ($f=a \times d \times e$). In this case it is $5 \times 1.25 \times 1.5 = 9.38$ or 9.4 rounded. The demand weight is simply the weight normalized so that the demand weight total equals 100.

The demand weight column shows that costs for service have the highest rating at 9.9, so this would be our highest design priority. Trade-in value is rated second at 6.3. Comfort in seating is rated third at 5.1.

People have asked me, "What happens if there's a tie in the demand weight?" Figure 5.16 depicts such a situation. The demand rates for features X and Y are the same at 7.4. A selection of X over Y means we wish to meet the customers need with a rating of 5 for X in our quality plan. A selection of Y over X means our quality plan rating of 5 for Y is driven by a desire to keep abreast of the competition. How does one resolve such a situation? You may try to give equal importance to both. But if your top-level priorities are worked out in advance, the answer is simple.

DEMANDED QUALITY DEPLOYMENT CHART

QUALITY PLAN

DEMANDED QUALITY				competitive analysis			plan			weight	
			importance (user input)	our company	company Y	company Z	quality plan	improve rate	sales point	weight	demand weight
1st level	2nd level	3rd level	(a)	(b)			(c)	(d)	(e)	(f)	(g)
cost	operation	service	5	4	4	4	5	1.25	◎	9.4	9.9
		fuel	3	4	5	4	5	1.25		3.8	4.0
		trade in	4	4	4	3	5	1.25	○	6.0	6.3
		-----	--	--	-	--	--	--		--	--
ease of use	comfort	seats	4	5	3	3	5	1.0	○	4.8	5.1
		heater	3	3	3	3	4	1.33		4.0	4.2
		noise	3	3	4	3	4	1.33		4.0	4.2
		dust free	2	3	3	3	4	1.33		2.7	2.9
		-----	--	--	--	--	--	--		--	--
capacity		-----	--	--	--	--	--	--		--	--
		-----	--	--	--	--	--	--		--	--
									total	94.5	100

d = c / b   e = 1.5 for ◎   e = 1.0 for no entry   f = a x d x e   g = f / f total x 100

e = 1.2 for ○

Figure 5.15. QFD Example for A Skip Loader.

QFD IS NOT A ROTE FORMULA -
AS IN ALL MATTERS, YOU MUST THINK ABOUT WHAT YOU ARE DOING

EXAMPLE: $a \times d \times e = f$

$5 \times 1 \times 1.5 = 7.5$

$3 \times 1.67 \times 1.5 = 7.5$

	competitive analysis			plan			weight	
importance	our company	company Y	company Z	quality plan	improve rate	sales point	weight	demand weight
(a)	(b)			(c)	(d)	(e)	(f)	(g)
X	5	5	4	4	5	1	◎	7.5	7.4
Y	3	3	5	5	5	1.67	◎	7.5	7.4
							⋛	⋛

X decision = meet customer need
Y decision = keep abreast of competition

A knowledge of priorities resolves such issues

Figure 5.16. A QFD Tie.

It should be emphasized that there is no single way to do QFD. Every company that does it does it somewhat differently. The example, while realistic, is meant to provide the general framework of how it works. Similar charts can be constructed as the product is broken down further for design, parts, cost, reliability, and the like.

As with all decision analysis techniques involving numerics, QFD begins with a subjective quantization of factors and ends up with what is perceived to be an objective finding. This is not always a bad idea, but it should be remembered that just because numbers are put on things (or printed out by a computer), it doesn't guarantee that the numbers are correct. In short, the whole field of multi-attribute decision analysis works only when the data are correct (of course the model has to be a good one, too). Still, the literature abounds with QFD success stories. The ones that aren't successful don't seem to get published.

Finally, with regard to priorities, there is increasing talk about how to manage cost-driven projects. I know of one company that assigned top managers to go off and study in detail how to manage them. They spent months on the task. The answer is quite simple. Think about it for a minute. Suppose you are in the business of instrumentation and you are asked to support the building of a platform to be used to gather science at an upcoming solar eclipse. A bunch of instruments of varying sophistication are called for. Science is the prime mission. But what is your highest priority? It is schedule. If you don't meet the schedule, the eclipse comes and goes and nothing happens. Everything else is subordinate to schedule. That is simple and clear.

In a cost-driven project, cost is the number-one priority and you have to be just as hard-nosed about it as you are about schedule for the eclipse platform. You set an envelope on cost at the outset. Everything else (schedule, performance, etc.) is subordinate. If the cost envelope is exceeded by one cent, you cancel. Quit. Give it up. You don't have to form an executive committee to figure out how to manage cost-driven product development. It's the same as any other, with cost as the hard-nosed, number-one priority. To succeed in a cost-driven environment, you have to know what to give up in advance to keep the project alive. That's why you need to define subordinate priorities.

People have said to me, "I understand what you're saying, but I don't think setting priorities will really work in our case." Wrong. Look at it this way: There will be a lot of decisions to be made during the actual development phase—a whole bunch by everyone involved. The decisions will be made on the basis of something. Count on it. Without priorities as guidelines, you run the risk of making decisions for one reason on Monday and for some other reason on Friday. Saying that you can't come up with priorities in advance is a form of denial or a kind of cop-out that says, "Can't I decide these things later on the fly?" Establishment of priorities in advance means that decisions will be made in a consistent manner on a consistent basis through time. This keeps everybody on the top-level track and saves a lot of time. What's wrong with that?

Establish your product priorities up front. They are a serious and valuable tool.

Planning for Reviews

This section discusses planning for reviews to be conducted during the product development phase. The purpose of a review is to

1. assure that all implementation activities continue to support the original customer needs

2. assure that all implementation activities are consistent with higher-level requirements

3. assure that implementation activities are meeting cost, schedule, functional performance requirements, and goals

The following paragraphs discuss the content and rationale for a basic set of formal and informal reviews.

A basic set of formal reviews consists of

- the product requirements review (PRR)
- the preliminary design review (PDR)
- the critical design review (CDR)
- the production part approval review (PPAR)
- the test readiness review (TRR)
- the pre-ship review (PSR)

A basic set of informal reviews consists of

- readings
- walkthroughs

This basic set will assist in your thinking. Other terms are often used, such as milestones or gates.

In any case, reviews are not held for the sake of reviews. They are held to gain visibility and to avoid surprises. If you don't need one, don't have one. On the other hand, you may want additional interim reviews as a function of perceived risk. For example, there may not only be a critical design review at the system, or total-product level, but there may be similar multiple reviews for selected subsystems as needed. An inherited engine may need little review except with regard to its interfacing with a new platform. A novel fuel injection, or braking system, may need its own set of complete periodic reviews. Reviews should not be punishing exercises. They are an opportunity to gain helpful input from peers. They are a solid part of the team effort.

Some top-level comments on the generic set of reviews follows.

The Product Requirements Review (PRR)

The PRR takes place following the formal development of product functional requirements. A basic purpose of the product requirements review is to demonstrate that the user or sponsor requirements are understood and that the basic requirements for meeting them are understood. This is accomplished by demonstrating a knowledge of *what* has to be done to meet the customers' needs and the identification of a feasible implementation approach (or sound implementation options), along with any attending risk and strategies for risk control. These considerations include functional requirements that map to the user needs, configuration management plans, interface documentation form and format, and any necessary trade-off analyses needed to support the feasibility of meeting stated requirements. The product requirements review may also cover the current status of resource plans and operating plans.

With passage of the product requirements review, the product development team earns the right to proceed with a preliminary design; that is, the project is allowed to finalize a design option and develop specifications. Upon successful completion of the product requirements review, the functional requirements document is signed off and is placed under configuration control. This level of design is referred to as the functional baseline. Procedures and plans for transition to the preliminary design review should also be covered at the product requirements review.

The Preliminary Design Review (PDR)

The purpose of the preliminary design review is to demonstrate the feasibility of the chosen design. At the preliminary design review, product specifications, test requirements, software requirements, the logistics support plans for development, vendor and contractor agreements, and interface requirements for the detailed design phase to follow are presented. The preliminary design review may also cover the current status of resource plans and operating plans.

Passage of the preliminary design review earns the right to proceed with a detailed design; that is, the product development team is then allowed to carry the design to a point of detail that would support initiation of actual fabrication.

The Critical Design Review (CDR)

The critical design review is appropriately named. The purpose of the critical design review is to demonstrate that the design is mature and ready for implementation. Hardware configuration item and computer program configuration item specifications, including schematics, engineering drawings, data flow diagrams, structured designs, pseudo-code, and so on, are reviewed. Logistics support designs, interface specifications, and test procedures are also reviewed. The critical design review may also cover the current status of resource plans and operating plans. The review is sufficiently complete so that implementation risk is well understood and acceptable.

Confidence of the implementors, sponsors, customers, and the review board must be high to pass the critical design review. Passage of the critical design review earns the right to actually build the system— that is, cut metal, initiate procurements, and begin formal coding.

In complex systems or products, there may be more than one critical design review. Programmatic or technical management may call for one such review for hardware and another for software, or may call for multiple reviews for elements or subsystems. A critical design review can last for hours or for days. Again, these determinations are totally dependent on management's level of confidence in design detail. As a rule, code production should not begin until completion of the critical design review. Exceptions to this rule may be tolerated. Some code production is required for early prototyping for the purposes of clarifying requirements or for justifying the realism of specific algorithms, and the like. However, this type of code production must not get out of hand, even if some of the work may be inherited into the final product. The perspective must always be maintained that prototype code supports the design process as needed and that implementation in earnest does not begin until successful completion of the critical design review. Software development planning should not assume that prototype code developed prior to the critical design review will be used during the later fabrication phase, which follows the critical design review. One of the toughest jobs facing the technical management is to keep design before implementation. This is particularly true in the software world.

The Production Part Approval Review (PPAR)

The purpose of this review is to demonstrate that supplier specifications, document control, and contracts are current and correct in support of the production part approval process. If you pass this review, you earn the right to go to production and assembly. This review is generally held in conjunction with the CDR.

Test Readiness Review (TRR)

A test readiness review may take place prior to the initiation of system integration and test, and following the development of complete test plans and procedures. Not all development efforts use test readiness reviews. In complex systems, or in situations where the desired visibility of technical management may be less than complete, test readiness reviews are highly recommended.

The purpose of the test readiness review is to assure that all plans and procedures are complete, all personnel understand their roles, and logistics support for testing operations is in place.

Testing—and, in particular, system integration testing—seldom, if ever, goes as planned. The course of testing is perhaps the most difficult enterprise to anticipate in any complex system development effort. Two items are commonly overlooked. One is the need for spares. This includes not only spares for the internal systems under test, but spares for the testing equipment as well. The second is adequate planning for regression testing with respect to when it is needed and its unanticipated effect on test schedules. Both of these items should be given particular attention at any test readiness review.

Pre-Ship Review (PSR)

PSRs are very appropriate for one-of-a-kind products. Such systems are seldom developed at the site where they are to be used. Moving a system from a development site to an operational site may be as simple as transporting a storage media or as complex as moving a massive collection of hardware components over a great distance and involving multiple modes of transportation.

The pre-ship review addresses two major questions:

1. Has the outcome of system-level testing been successful, and as a corollary, is the system ready for customer acceptance testing?
2. Is the required logistics support for system transfer, installation, and customer acceptance testing well designed and ready to set into motion?

System-level tests are different from system-integration tests. SI&T deals with the methodic sequential assembly and testing of internal system interfaces and system-level interfaces. The emphasis of system testing is on system-level responses to system-level inputs, that is, the system that the customer sees. Pre-ship system-level tests should employ an exact replica of the operational setting whenever feasible. This includes the same placement of equipment, use of cables, environment, and so on. Simulations of the user environment, when necessary, should be noted and reported on as a part of the review.

Pre-ship system-level tests are often more elaborate than final customer acceptance tests. The pre-ship review thus addresses both issues of the system's ability to meet the full set of original functional requirements, and the system's ability to pass customer acceptance testing.

The pre-ship review also addresses the adequacy of the complete logistics support design to accomplish transfer, installation, and acceptance testing. This includes issues related to supply of all materials, test equipment, transportation and handling, technical data packages, required facilities, personnel/training, and system maintenance plans.

Formal Review Personnel

Personnel associated with formal reviews consist of the convening authority, a review board chairperson, review board members, a responsible presentation manager, and a review board secretary. The review board members typically sit at a table at the front of the room facing the presenter(s). An audience of all interested parties is accommodated in the remainder of the room.

The convening authority is the individual who calls the review in accordance with the project schedule. He or she designates a review board chairperson, approves the purpose for each review, and approves the composition and members of the review board.

The review board chairperson selects the review board members and a review board secretary. The chairperson designates a responsible manager to present the review material and works with the designee to construct an agenda. The chairperson manages the review and also assures that all requests for action generated during a review are appropriately addressed on or before assigned due dates. Requests for action may be written up and submitted on RFA forms during the progress of the review or during a standard polling of the board by the convening authority at the end of the review.

The review board should consist of eight to ten peer experts covering the programmatic, process, and technological issues to be reviewed. All formal review boards should always include customer representation.

The responsible presentation manager is usually a project manager, deputy project manager, or the product development team leader. The presentation manager may call upon any supporting personnel to give appropriate portions of the presentation. He or she also assures that copies of the review presentation are circulated to board members at least one week in advance of the review and arranges for all other required support (facilities, presentation materials, refreshments and so on).

The review board secretary is responsible for taking minutes of all proceedings, with special attention toward coordination and documentation of requests for action. Following the review, the secretary generates a summary board report in memo form, including salient findings, for concurrence by the presentation manager and chairperson. The secretary also maintains a file of all review board actions.

Informal Readings

Readings are conducted by one or more selected peers to provide early exposure on documentation progress. Peers don't have to be on the product development team. They are typically other cognizant engineers or staff members familiar with the project. Readings should be routinely requested by all authors. Documents should be reviewed for consistency in form and format, as well as for the integrity and completeness of their content. In particular, repetition of material in more than one document should be avoided to alleviate excessive work associated with change control.

Comments offered by peers should be clear and useful. For example, simple comments such as "not clear" or "recast" should be avoided in favor of actual suggested text replacements or clear suggestions as to how the material may be improved. Incorporation by the author of comments and suggestions made by peers is not mandatory, although serious consideration should always be given to any opportunity to improve communications.

Documentation includes software pseudo-code and code listings. An extremely effective method for checking code is simple peer review through reading. This is accomplished by asking programmers on the project to set time aside to read each other's code in isolation. It is often difficult when one is involved at levels of considerable detail to see simple errors, or to provide adequate comments, when dealing with material that is familiar and seems obvious to the primary programmer. Peer review through simple reading provides an opportunity for a truly fresh look at the work. Finding errors in code through peer review is especially valuable in that a relatively small investment of time at this level can save significant amounts of time later if the errors need to be corrected at the integration and system test levels of effort.

Informal Walkthroughs

The walkthrough is an in-depth technical review of the progress of hardware or software requirements development, design, and/or implementation. Walkthroughs are basically mini-reviews conducted on specific parts of the overall system design. Their frequency and timing is solely up to the development team leader and is determined by the need for visibility. Schedules for walkthroughs are typically made to accommodate the concurrent product development team membership. Additional walkthroughs may be planned at any time.

On occasion, the walkthrough participants may consist of only those team members who have direct interfaces with the subsystem in question. These include the team leader, cognizant hardware and/or software engineers, and selected peers with direct interfaces with, or knowledge of, the work under review. Presentation material should be standardized and distributed to attendees in sufficient time for review prior to the walkthrough.

It should be emphasized that a walkthrough is not a performance evaluation. It is actually an opportunity for the reviewee to greatly enhance the probability of his or her success in dealing with the complete scope of implementation issues. The value of walkthroughs is often degraded by the presence of upper managers, because a different tone can be set in which criticism may carry a connotation of ineptness. Walkthroughs are seldom of any value unless there is a complete sense of freedom and a common goal of identifying problems and potential problems, and working together to correct them. The sessions must be wide-open, and are most valuable when no attendees are threatened and all are of a common, problem-oriented mind-set. The product development team leader must clearly set this tone.

Walkthroughs can be held to review requirements, design, code, test, or at any other point considered necessary. The following paragraphs cover the major considerations to be addressed at requirements, design, and code walkthroughs.

Requirements Walkthrough Topics
- A statement of functional requirements in the context of total system partitioning.
- An English-language summary of derived requirements.

- A listing of derived requirements with traceability of each to the system functional requirements by paragraph number.
- A clear description of interface requirements.
- Are the requirements achievable both technically and programmatically?
- Are the requirements testable, and what method will be used for their testing?

Design Walkthrough Topics
- A complete design description using approved structured tools such as CAD, structured programming, and so on.
- The ability of algorithms to meet required functions and computational error envelopes.
- Is the design traceable to the requirements?
- A description of all designs including drawings, schematics, data structures, databases, and so on.
- A description of all error-handling routines and mechanisms.
- A clear description of interface design.
- A report on margin status as appropriate.

Software Code Walkthrough Topics
- Does the code comply with the design?
- Are comments sufficient for the unfamiliar reader?
- Is the language efficiently and properly used?
- Are data constants, typings, and declarations correct?
- Is the design documentation updated and complete to date?

Planning for Reporting

The purpose of a report is to assist programmatic and/or technical managers in the assessment of the status of product development progress.

This section discusses planning for the form, format, and content to be used in support of periodic reports made to management at all levels as a part of the evaluation process. The term "report" is also a noun used to denote documentation such as reports of formal review results, testing results, concurrent product development team products such as trade-off analyses, risk analyses, design team minutes, and so on. In the meantime, it should not be confusing to recognize that the reports discussed in this section also wind up as documents.

The following paragraphs discuss the content and rationale for a basic set of reports for use in reviews and other periodic evaluations. A basic set of formal reports consists of

- executive summaries
- schedule reports
- cost reports
- technical process reports

Before discussing these formal reports in detail, let's look at a number of characteristics that are common to all reports. All reports should provide information on

1. comparisons of current actual values with:
   a. the current established project baseline
   b. values presented in previous reports
2. forecasted trends
3. identification of problem areas
4. proposed actions over the next reporting period

Information presentation should focus on the use of charts and tables, with supporting explanatory text as required. Charts are used to represent progress against a time base. Crisp notes are included near the bottom of each chart to highlight salient points. Charts should include a sufficient number of previous and future reporting periods to adequately show past history and forecasted data. It is useful to maintain the current report period near the middle of the span of reporting periods presented so that actuals, trends, and predictions are clearly displayed. Tables are used to support the charts by providing additional numeric detail for clarification as required.

All report metrics, units of measure, and forecasting techniques should be standardized across all report types and reporting levels to the extent possible. Standardization, however, should not be imposed to such an extent that clarity of information transfer is compromised. Deviations should be allowed in the interest of clarity. The need for communication wins out over the need to obey rules. The advantage of some modicum of uniformity in report formats and structure, particularly on large projects, should be obvious.

Responsibility for generation and presentation of reports follows the generic WBS. Report data at a given WBS level are produced by product-level managers and by concurrent product development team members at that level and presented to, or provided to, management at that level. These reports feed into and are summarized at the next highest product level of the WBS. For example, in the automobile WBS shown in Figures 5.2 through 5.7, the chassis manager accumulates report data for each of his or her products (steering, brakes, tires/wheels, etc.) and reports to the vehicle manager. Similarly, the engine manager receives reports from the person or persons responsible for the cylinder block, cylinder head, valve train, and so forth, and reports to the power train manager. That is, the reporting structure and information covered are determined *exactly* by the structure of the WBS. In this manner, report data are generated, increasingly summarized, and presented bottom-up. This approach allows conformance to baseline plans at lower levels to be routinely absorbed into data presented at higher levels. Deviations at lower levels, however, tend to emerge upward to provide needed visibility at higher levels.

To assure that information is not lost through the normal aggregation process, all reports at all levels should be generated using an integrated computerized system and maintained in a single electronic repository. While report data may be aggregated and summarized at each higher level, each report exists as an accessible stand-alone document at a given WBS management level. This approach preserves vital information at each WBS level and assures that lower-level status information is not hidden by the aggregation process. For example, a cost overrun for a product at level N may appear at that level while the product costs at the next highest level may not appear as an overrun. This may occur if a balancing cost underrun is taking place at one level such that costs are in conformance at the next highest level. The fact that there is an overrun at level N, however, is preserved because reports remain intact for each level.

This reporting structure is a model designed to promote efficiency and completeness in information transfer from the bottom up, yet to maintain important information at all levels for review at any time.

A major advantage of this approach is that *all reports at all levels remain easily accessible for review by any worker or manager at any level at any time through a single electronic resource.*

Executive Summary Report Content

The executive summary report provides top-level information on the current reporting period status of costs, schedules, and technical process as it relates to the current approved baseline. The report uses green, yellow, and red indicators of status. A sample executive summary chart is shown in Figure 5.17 The example shows an executive

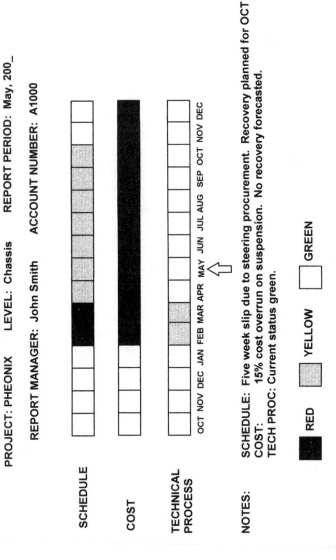

PROJECT: PHEONIX   LEVEL: Chassis   REPORT PERIOD: May, 200_

REPORT MANAGER: John Smith   ACCOUNT NUMBER: A1000

SCHEDULE

COST

TECHNICAL
PROCESS

OCT NOV DEC JAN FEB MAR APR MAY JUN JUL AUG SEP OCT NOV DEC

NOTES: SCHEDULE: Five week slip due to steering procurement. Recovery planned for OCT
        COST: 15% cost overrun on suspension. No recovery forecasted.
        TECH PROC: Current status green.

RED   YELLOW   GREEN

Figure 5.17. Executive Summary Report Example.

summary report generated by the Phoenix chassis level manager. A guideline for the use of indicators is as follows;

- GREEN   Zero or small deviation from current baseline.
- YELLOW   Significant deviation from current baseline, but a low-risk plan for recovery is in place.
- RED   Significant deviation from current baseline and a medium- to high-risk plan for recovery is in place, or a plan for recovery is in development, or no recovery is forecasted.

The executive summary report may be used at any WBS management level. Thresholds and/or values for indicators should be clearly noted if different from an established programwide set. Information is presented for previous report periods, the current period, and the forecasted period.

In the example of Figure 5.17, there was a five-week delay in steering procurement starting in February. This caused a red condition until April, when recovery plans were completed. Complete schedule recovery is expected by October. Until such time, the schedule yellow flag remains up. In February, a 15 percent cost overrun on the suspension occurred. No recovery is forecasted. Technical progress was hindered by the steering delay in February and March, but a complete catch-up was accomplished by April. Note that the current reporting period of May is indicated by the arrow.

Schedule Report Content

Schedule reports provide information on the time status of all products. The information includes timing for current baseline product deliveries, operating plan product deliveries, and current actual estimates for product deliveries if slippages to the baseline or operating plan are anticipated. Reasons and impacts of deviations may be presented in bullet form using separate supporting text.

Products include management products, concurrent product development team products, the mission products, logistics support products, and test products. Products also include hardware, software, negotiated internal system receivables and deliverables, and *all* documentation (reports, review reports, requirements, specifications, test plans and procedures, test reports, and so on). All products exist on the WBS.

Gantt charts provide a convenient format for presenting schedule progress on a time line. Items included in scheduling reports should conform to the work flow diagrams for the level being reported on. The work flow diagram, in turn, is consistent with the WBS.

An example of a Gantt chart schedule is given in Figure 5.18. Most organizations have their own internal versions of Gantt charts. In the example, the status of each WBS product is shown for a reporting period of May 200_. The time lines indicate the planned work periods required to support production of the products. A further breakdown of product scheduling at level N is presented in the Gantt chart at level N + 1, that is, the next level down. In the example, the five-week steering slip reported by the chassis manager is indicated with recovery in October. A more detailed accounting would be presented at the next level down. The chart also shows the history of a seven-week slip in production of the logistics support plan.

Note that the items scheduled in the sample Gantt chart are products. There has been long-standing and repeated debate in many organizations as to whether schedules should show events or products. The discussion invariably centers around a concept of unknown origin that one *or* the other should be shown, presumably in the interest of some rule of consistency. The fact of the matter is that the purpose of any report is to convey information, not suppress it. Should it be desirable to indicate the timing of events, such as a preliminary design review, and so on, there is really no reason why the planned occurrence of the event cannot be shown as a reference point on a Gantt chart for clarity. But bear in mind that a design review also has a product associated with it, namely a design review report. Gantt charts should always be primarily product oriented and include every product. This is because the product-oriented mind-set maintains clear accountability at every level. Events associated with timing are items such as initial operational capability, delivery dates, initiation of integration and test, and so on. Events may be incorporated as useful reference points, but do not inherently indicate accountability. Thus, Gantt charts should *always* show products at each level and *may* indicate event timing as reference points only.

A few more words on the word "product." There is an end product and there are intermediate products needed to create the end product. An end product is meant to be bought or accepted by a final user or set of users—generally the consumer. An intermediate product is created by the developer as a result of a well-defined process carried out in direct support of the development of the end product. Intermediate products consist of documents or physical entities. Thus, items such as requirements, specifications, reports, test rigs, prototypes, subsystems, assemblies, trade-off analysis, team minutes, and reviews are examples of intermediate products.

PROJECT: PHEONIX   LEVEL: Chassis   REPORT PERIOD: May, 200_

REPORT MANAGER: John Smith   ACCOUNT NUMBER: A1000

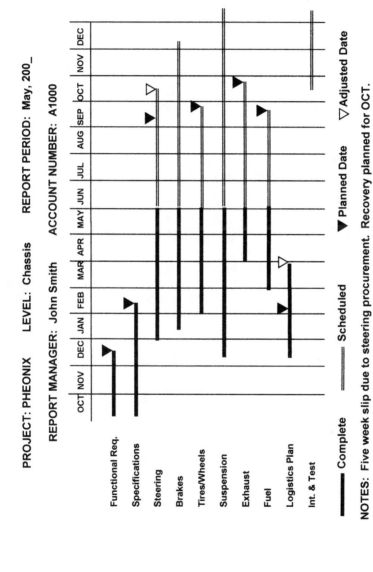

NOTES: Five week slip due to steering procurement. Recovery planned for OCT.

━━ Complete   ═══ Scheduled   ▼ Planned Date   ▽ Adjusted Date

Figure 5.18.   Gantt Chart Report Example.

This leads to an important distinction to be made between a product and a process. A process is a procedure, or a set of steps to be followed, for the specific purpose of developing a product, whether it be intermediate or final. Product development should not begin solely with the identification of processes, for it is easy to develop non–value-added processes that are either unneeded and/or do not result in specific products.

Rather, product development should entail the identification of the required set, *and only that set,* of intermediate products necessary and sufficient for the development of the end product. Processes may then be developed to achieve realization of products, but it is not the other way around. That is, we are not process oriented, inventing processes for the sake of processes. We are product oriented, and all processes we devise and carry out are specifically for the purpose of realizing products alone. Our WBS is made up of products, and it is the status of those products that we report on in our schedule reports.

Cost Report Content

A cost report example is shown in Figure 5.19. In this example, planned, actual, and forecasted costs are displayed for the Phoenix chassis. A 15 percent overrun for the suspension element is shown, with no recovery anticipated or planned for at the current reporting period. The exact nature of the cost overrun would be indicated in the suspension element cost report at the next level down. Costs include the following four basic components:

- workforce, including contractors and consultants
- support services such as graphics, travel, and so on
- procurements
- overhead

Detail on these cost items may use supplemental charts using the same format example shown in Figure 5.19. This format may also be used to show the history, current status, and future projections for cost reserves and obligations. Reserves are contingency funds. Obligations include known costs that will be incurred at a future time and which have not been paid to date.

While the schedule and cost charts shown thus far are useful, they do not always clearly correlate the progress of work with its costs. A more sophisticated approach designed to clarify the relationship between expenditures and the actual progress of work at any given point in time is offered by the cost, schedule, and performance tracking concept—commonly referred to as earned value. In this scheme, variances

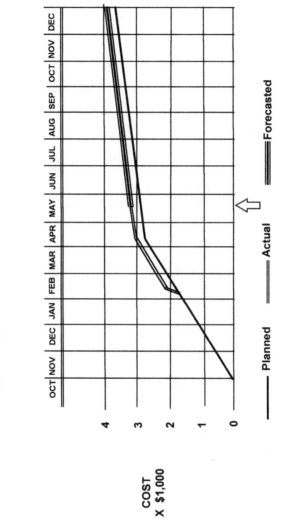

PROJECT: PHEONIX   LEVEL: Chassis   REPORT PERIOD: May, 200_

REPORT MANAGER: John Smith   ACCOUNT NUMBER: A1000

NOTES: COST: 15% cost overrun on suspension. No recovery forecasted.

Figure 5.19.   Cost Summary Report Example.

157

in both cost and schedule are measured in terms of planned and actual expenditures. Three new terms are introduced for these purposes:

BCWS   The budgeted cost of work scheduled (BCWS) is the estimated cost to perform the required work based on the original project schedule

BCWP   The budgeted cost of work performed (BCWP) is the original budgeted estimate of the actual worked performed to date

ACWP   The actual cost of work performed (ACWP) is the real cost of the actual work performed to date

Each of these terms has its units in dollars and can be plotted as a function of time. Figure 5.20 gives an example of how these concepts interrelate. The single line in the figure, which is the BCWS, shows the profile of the cost as originally planned. The double line tracks the actual cost incurred for the actual work performed up to the reporting date. The double line to the right of the reporting date displays the estimated cost to complete. The lower triple line shows the original budget that was allocated for the actual work completed at the report date.

In this formulation of data, the difference between the BCWP and the BCWS can be viewed as a measure of schedule variance. In the example of Figure 5.20, the schedule variance suggests that the work is behind schedule because the budgeted cost of the work performed is less than the budgeted cost of the work scheduled to be completed at the reporting time. There is also a cost overrun because the actual cost of work performed is higher than the amount budgeted. These discrepancies in terms of schedule and cost variance are added to the ACWP curve to arrive at a new estimate of schedule completion and required costs. The approach assumes that the BCWP is a faithful measure of the actual work performed. When this is true, the concept can provide a single performance measurement framework in which to view the relationships between cost and work completion.

Technical Process Reports
Technical process reports include all other planned reports other than executive summaries, schedule, and cost reports. These include reports that cover

- past, present, and projected data on the margin management of dear commodities such as power, weight, response time, and so on

- past, present, and projected data on configuration management items such as engineering change requests, failure reports, technical liens and waivers, and action items

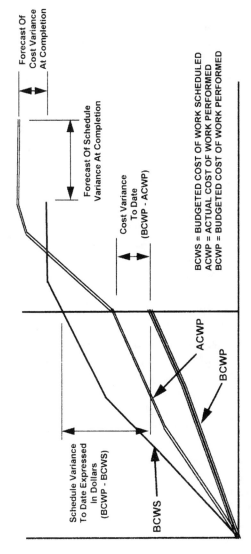

Figure 5.20. Earned Value Concept.

Forecast Of Cost Variance At Completion

Forecast Of Schedule Variance At Completion

Cost Variance To Date (BCWP - ACWP)

Schedule Variance To Date Expressed In Dollars (BCWP - BCWS)

ACWP

BCWP

BCWS

BCWS = BUDGETED COST OF WORK SCHEDULED
ACWP = ACTUAL COST OF WORK PERFORMED
BCWP = BUDGETED COST OF WORK PERFORMED

159

An example of a weight margin report is presented in Figure 5.21. The planned, or allocated, weight for the chassis of the Phoenix is used in the example and is shown as a solid line. The actual best estimate of weight as a function of the current design at the time of the reporting period is shown as a double line. The current margin at the reporting month of May is 150 pounds. The forecasted consumption of the weight margin is also displayed as the triple line. A final margin of 100 pounds is estimated at the time of delivery of the Phoenix vehicle.

Similar formats may be used to report on the status of other dear commodities identified during the development of technical controls and policies. Tables may also be used to provide planned, current best estimate, and forecasted values in numeric form for each reporting period.

The trend format is also useful for reporting on the status of engineering change requests, failure reports, liens and waivers, action items, and so on. Figure 5.22 shows such a chart for action items associated with the Phoenix chassis. Since technical processes of this nature are unplanned, there are no forecasted values presented in these reports. As in the case of margin reports, tables may also be used to provide numeric details. In this case, tables would show accumulated data for each reporting period up to and including the current one. Trend reports in graphical form are of particular value in that they immediately convey a sense of the degree of control that is being exercised over both planned and unplanned events.

Planning for Margin Management

Planning for technical margin management involves the allocation of sufficient contingency for commodity budgets across segments, elements, and subsystems and the identification of processes for margin management throughout system development. For example, any flight system will always involve the establishment of commodity budgets for overall mass, power, center of gravity, and payload weight and volume, in addition to other items deemed critical to the success of the mission design. Earth-based communications systems typically involve budgeting of resources such as overall system throughput, response times for specific message types, memory, error rates, system availability, and so on.

The setting and managing of margins entails

1. Determination of product performance characteristics that are desired for the product baseline that is to be delivered to the user

2. Determination of initial margins to be applied to achieve product baseline goals and to assure adequate margin throughout the uncertainties of design, fabrication, and test

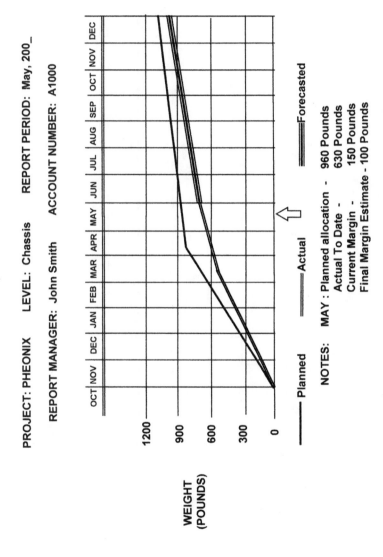

Figure 5.21. Weight Margin Report Example.

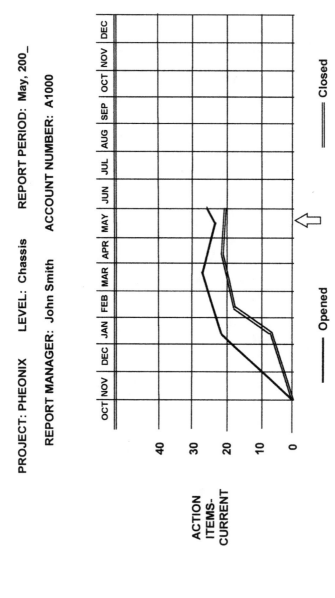

Figure 5.22. Action Item Report Example.

162

3. Development of a plan to conservatively allocate margins across internal systems and to periodically monitor the disappearance of allocated margins

Note that, ideally, margins devoted to the development process should methodically dwindle to the target margin for the product baseline. The target margin may or may not be zero. The situation is very similar to planning for, and using up, financial contingency. If a project comes to conclusion without expenditure of planned financial reserve, or if a cost overrun is experienced, the project may be subject to criticism regarding the management of monetary margin, or reserves. The terms "budget management" (i.e., dollars) and "technical commodity margin management" are interchangeable in this sense. Thus, the setting of margins for dear commodities and their allocations across segments, elements, and subsystems requires the same conceptual mix of aggressiveness and conservatism that financial planning does. Typical resources that often require budgeting are

Mass

Power

Computer memory

Response time

CPU utilization

Bandwidth

System availability

System reliability

Error rates

Target circular error probability

I/O ports

Volume

Size

Payload

Thermal

Vibration

Expendables

Survivability

All the other "ilities"

The list would include anything else that technical management believes should be monitored.

❒⁝ **TABLE 5.9**   *Guidelines for Hardware-Related Margins*

For	Margin in Percent
Guesses and "great ideas"	20
Sketches, schematics, and drawings	10
Off-the-shelf known items	2
Exact replicas	0.5

The values discussed below are given as guidelines only. The actual values used for both hardware and software will depend on the level of confidence of the team leader and on the assistance derived from the concurrent product development team members. The use of outside experts is also encouraged in this effort. It is always preferable to gather values based on experience from similar projects. If empirical data cannot be acquired, it is wise to widen margins as much as possible while still meeting system end-to-end performance requirements.

Typical hardware margin values are given in Table 5.9. The widest margin is given to guesses and "great ideas." Clearly, this is the category in which the greatest risk lies and 20 percent probably represents a lower limit. Sketches, schematics and drawings, generally enable the making of some kind of assessment as to part types, part counts, required power, circuit performance, availability, and the like. If these analyses are carried out with care, and confidence in their findings is good, a 10 percent margin should be adequate. If not, scale up. When dealing with off-the-shelf known components that are mature and have believable specification sheets, or if knowledge is based on reliable experience, a 2 percent margin should be reasonable.

Occasionally, a system segment, element or subsystem may consist of an exact replica of one previously employed. If it is felt that there is little chance that the new application may require unforeseen changes, the margin on commodities associated with the replica could be as little as 0.5 percent.

A starting point for determining software-related margins is given in Table 5.10. Computer memory is a commodity that is almost always consumed at a rate higher than anticipated. When engaged in building a system very similar to one that has been built before, or if good documentation on a very similar system is available, one may start with 100 percent memory margin with a 25 percent margin as a product baseline goal. If there is any question at all about this, use a much bigger margin. Because memory is cheap and

█ TABLE 5.10 *Guidelines for Software-Related Margins*

For	Margin in Percent
Memory	At least 100% to start, wind up with 25% unused
CPU utilization	Less than 70% with random arrivals
Response time	10% to 100%
(Note: Software requirements typically drive hardware requirements.)	

requires relatively little space, weight, and power, and the like, use of a more substantial margin is often possible—often up to 500 percent. Make estimates for all software involved. Estimates are generally easier for existing operating systems, compilers, utilities, and so forth. The big unknowns lie in new executives and application software. Break these down as much as you can at this early stage, making estimates for each program even though the design has not really begun. Clearly at this stage you must err on the conservative side.

After delivery, software systems undergo software maintenance that invariably calls for periodic operating system upgrades, general software package upgrades, corrections due to failure reports, and enhancements to application software resulting from routine change requests. Considerations for memory management should address these additional memory needs, in addition to any additional uncertainties involved with the development cycle.

Computer processing unit (CPU) utilizations must never exceed 70 percent with random arrivals for requests for service. This means, if a 100 percent margin is set at the outset, the initial design will call for a 35 percent CPU utilization. CPU utilization is a very important commodity to plan for and monitor. There are a lot of systems in service that run too slowly under peak average loads, or meet requirements at installation time but rapidly deteriorate with reasonable growths over the systems lifetime.

Meeting end-to-end system response time goals also typically requires application of conservative margins. Again, the margin assigned depends solely on the level of faith that requirements can be met. Should there be any serious reservations, try to start out with a 100 percent margin.

In dealing with computers, it is almost always best to begin by setting software margins. Stated in alternate terms, computer hardware requirements should be driven by computer software requirements. That is, the setting of computer software margins is the starting point for the setting of computer hardware margins. There are rare cases in which it can happen the other way around, such as when you are constrained by the use of inherited hardware for an upgraded or new system, but it is seldom desirable to start this way.

Many military systems call for preplanned product improvements—or so called P³I. Some commercial products are also purposely designed to accommodate upgrades. These product enhancements are improvements that are planned for as a part of the present design and development process for anticipated improved versions of the product to be implemented, or added at a later time. These considerations, in effect, call for placement of higher levels on a pertinent set of your margins. This, in itself, is no small exercise and must be considered in these early stages of setting margin values.

Actual estimates will, of course, vary around these guidelines. The margin values selected are a direct consequence of the level of certainty associated with the desired targets. Ideally, margins are determined from solid experience. When this is not possible, the expertise on the product development team should be called upon to perform analysis, simulation, or even preliminary bread boarding or brass boarding. The setting of margins is also a valid point for the concurrent product development team to call upon outside experts as required.

Planning for Documentation

A document is a product that is a textual or graphic description of a product or process. Documents may be in hard copy or electronic form.

Documents are created to support the system development process and to support system operations. The purpose of documents created to support the system development process is

1. to foster communication between all development team members by providing permanent sharable records
2. to provide a basis for review of the status of development products

The purpose of documentation to support system operations is to provide information on the proper use and logistics support of a system in its operational setting.

Documentation Structure

Documentation structures are often represented as tree structures. Whereas the use of tree structures is widespread, it should be emphasized that the true framework for the structuring of documentation is the work breakdown structure. Documentation should be assigned to appropriate work breakdown structure items and at appropriate levels. Table 5.11 gives examples of document subject assignments to the generic work breakdown structure.

The table shows basic subject matter for each work breakdown structure category. The exact content of a given document may vary at different levels. For example, the system level typically produces top-level functional requirements, while concurrent product development team members at the segment, element, and subsystem levels produce successively more detailed levels of design in response to higher-level requirements.

Statements Used in Documents

Documents may be viewed as containing four types of statements: requirements, guidelines, statements of intent, or information.

A requirement is a statement about a product characteristic that must be satisfied, or an activity that must be carried out and is characterized by the verb *shall*.

There are two types of guidelines: recommendations and options. A recommendation is a guideline for a product or process that carries the strength of a desire and is characterized by the verb *should*. An option is a guideline for a product or process that carries the strength of an alternative and is characterized by the verb *may*.

A statement of intent describes a nonbinding future action concerning a product or activity and is characterized by the verb *will*.

A statement of information is any statement that is not a requirement, guideline, or statement of intent.

Document Types

A document may be viewed as belonging to one of six types. A convenient classification for document types and their uses is presented in Table 5.12.

Document States

Document versions should exist in well-defined states. A useful set of document states and definitions is presented in Table 5.13.

▯ TABLE 5.11 Examples of Document Subject Assignments to the Work Breakdown Structure

WBS Item	Subject
Management	Preproject plan
	Project plan
	User needs
	Work breakdown structure
	Organization
	Control policies/plans
	Resource plan
	Segment, element, etc., plans
	Reports
Concurrent Development Team	Concurrent product development team management plans
	Requirements
	Specifications
	Designs
	Interfaces
	Test requirements
	Trade-off studies
	Reports
Logistics Support	Logistics supply plans
	Supply
	Test equipment
	Transportation
	Facilities
	Training
	Maintenance
	Reports
Testing	Test plans and procedures
	Parts testing
	Unit testing
	Integration testing
	System testing
	Acceptance testing
	Reports

▯⋮ TABLE 5.12 *Document Types and Uses*

Type	Use
Standard	Used to promote consistency or to promote good practices
Handbook	Used to promote good practices and provide useful information
Specification	Used to describe requirements for a specific product or process
Technical Manual	Used to describe how to use, operate, or maintain a specific product, or how to perform a specific process
Plan	Used to describe the approach and practices to be followed to make, acquire, test, or use a product or process
Report	Used to assist programmatic and/or technical managers in the assessment of the status of system development progress

▯⋮ TABLE 5.13 *Document States and Definitions*

State	Definition
Planned or Identified	A document for which a number has been assigned
Draft/In Work	A document on which work has begun
Preliminary	A document that is ready for internal review
Final	A document ready for approval signatures
Approved	A document that has signatures of approval and concurrence as required
Released	A document that has been approved and is formally available to document users
Revised	A document that has been approved or released and has been altered

Documents should be approved by the management at the work breakdown structure level at which the document was produced. Concurrence may be obtained at higher work breakdown structure levels as desired. There should be no formal change control on documents until they are approved. Documents in any state should be available to all development team members. Development team members should include representatives of the user community (e.g., the customer) and the supplier community. Documents are made available to the complete user community upon being released.

Document Parameters

An electronic document database should be maintained to allow remote access to all documents from all development team work areas and facilities. The database should include the following document parameters as a minimum:

- document title
- document number based on the work breakdown structure category and level
- document state
- date of document state
- document type
- author(s) by job title consistent with the work breakdown structure
- approval(s) and concurrence(s) by job title consistent with the work breakdown structure

Finally, documentation should always be well planned and limited to the minimal set required, but not to the extent that efficient information sharing is compromised. Examples of factors that influence the amount of documentation required are: the number of mission product items, development team stability, development team experience, the physical location(s) of the development team, the multiplicity of disciplines involved, and the need to maintain an operational system over a long period of time.

Documentation at lower levels should always avoid repetition of material presented at higher levels, even if it means that some resulting documents may contain multiple references and be physically small. Documents do not have to be big. This guideline potentially alleviates the complexity of the configuration management of document contents.

Occasionally, it may be desirable to construct stand-alone documents for procurement purposes. Stand-alone requests for proposals, or requests for quotations, should be constructed by assembling needed portions from the set of development documents over the affected levels. For example, a procurement request at the subsystem level (a sensor, or a navigation platform, or a steering system, etc.) may contain portions of the system-level functional requirements (developed by the concurrent product development team at the system level) to provide a system overview, and a set of detailed design requirements developed by the concurrent product development team at the subsystem level.

Planning for Risk Control

Planning for risk control involves the identification and quantification of technical risk and the formulation of alternative backup strategies when required.

In its simplest form, the steps in planning for risk control are

1. Identification of the nature of the risk. Is it associated with schedule, cost, technical issues, or one or more of each?
2. Development in advance of one or more strategies to remove or minimize each risk.
3. Determination of a given point in time during the product development cycle that backup strategies will be put in place, if needed.

Thus, if a given technology is risky at the outset, an alternative low-risk technology should be identified during Phase B and a specific time to revert to its use during Phase C should be decided upon. These considerations generally impact cost and schedule considerations. Therefore, it is best to understand and build in these factors up front.

There have been volumes written on the subject of quantifying levels of risk. Although it is beyond the scope of this book to cover this issue in detail, I include one method to provide the flavor of the thought process and of the potential errors inherent in such approaches.

One common method of quantifying risk involves the making of estimates for the probability of failure, P(f), and the consequence of failure, C(f).[2] These quantities are then used to calculate an overall Risk Factor, RF, which is a measure of the likelihood of failure.

The value P(f) is a measure of the lack of success of a given system, subsystem, or lesser part of a system. Typical considerations are the maturity and complexity of hardware and software. The approach also allows for inclusion of other factors. The equation for P(f) may be expressed as

$$P(f) = (P_{MH} + P_{MS} + P_{CH} + P_{CS} + P_D) / 5$$

where

P_{MH} = Probability of failure due to hardware maturity,

P_{MS} = Probability of failure due to software maturity,

P_{CH} = Probability of failure due to hardware complexity,

P_{CS} = Probability of failure due to software complexity, and

P_D = Probability of failure due to other factors.

The magnitude of the term in the denominator is adjusted to equal the number of terms in the numerator.

The value C(f) is a measure of the effect of failure of an element. Typical consequences of failure can impact technical, cost, and schedule factors. The equation for C(f) may be expressed as

$$C(f) = (C_T + C_C + C_S) / 3$$

where

C_T = Consequence of failure due to technical factors,

C_C = Consequence of failure due to cost factors, and

C_S = Consequence of failure due to schedule factors.

The risk factor, RF, is the overall likelihood of failure and is expressed as

$$RF = 1 - [1 - P(f)] * [1 - C(f)]$$

Table 5.14 lists sample rationale for the assignment of values for the components of probability of failure and consequence of failure for a system that will use existing computer hardware and software with the exception of minor changes in displays. We assume the system in question also involves research and development for a novel high-performance transducer that is reduced in size and designed to replace an existing bulky technology. Although the transducer concept is not mature, it is considered promising in that it is not, in itself, complex.

The values for P(f) and C(f) are calculated as follows

$$P(f) = .32$$

and

$$C(f) = .50$$

▮⋮ **TABLE 5.14**   *Sample Assignment and Rationale for Components of P(f) and C(f)*

Component	Value	Rationale
P_{MH}	.8	Transducer concept is new, research is required.
P_{MS}	.3	Existing software data processing package will be used with minor display modifications.
P_{CH}	.2	Computer hardware inherited, transducer concept involves low complexity.
P_{CS}	.2	Low complexity in terms of number of software modules.
P_D	.1	Negligible risk associated with vendors and subcontractors.
C_T	.7	Significant degradation in performance if novel transducer concept fails and backup strategy to existing technology is required.
C_C	.5	New transducer proof of concept expected early—moderate cost risks.
C_S	.3	If new transducer concept not verified on schedule, revert to existing technology—minor schedule impact anticipated.

The overall risk factor, RF, is then calculated as

$$RF = 1 - (1 - .32)(1 - .50) = .66$$

Table 5.15 provides a guideline for the assessment of the risk factor RF. In the example given, the risk factor is considered to be moderate to significant. While such rating factors can be of value in attempting to bring quantization to the degree of anticipated risk, it should be noted that rating schemes are among the more simplistic approaches within the broader discipline of multi-attribute decision analysis.[3] Also note, as with QFD and other techniques that bring comfort in quantization of outcome, the original assignment of numerics to variables and parameters is of extreme importance. Garbage in, garbage out, as we all know. Be particularly wary when the outcome values, upon which decisions

◫ **TABLE 5.15** *Risk Factor Magnitude Guidelines*

Magnitude	Interpretation
.1	Negligible
.2	Low
.3	Minor
.4	Minor to moderate
.5	Moderate
.6	Moderate to significant
.7	Significant
.8	High
.9	Extreme

will be made, are numerically close. Check the sensitivity of outputs on minor changes of inputs. Before you rely on such techniques, be sure you have something that is meaningful.

Planning for Configuration Management

Configuration management is the discipline of identifying selected product, or system, items at discrete points in time for the purposes of systematically controlling changes and maintaining the integrity and traceability of the configuration throughout the development cycle. Candidate items for which we may wish to maintain change control are the product and its segments, elements, subsystems, components, software, documentation, and the like. Selected items are referred to as configuration items (CIs).

This section presents basic definitions and processes that should be included in planning for configuration management. The material is oriented around classic configuration management associated with use of the staircase system development paradigm. When the spiral model or the rapid development method for system development is used, modifications are required to serve the needs of flexibility as to how and when different levels of control are applied to these paradigms. In the early prototyping and spiral models, the modifications consist of an extension of the early, more relaxed phases of traditional configuration management with regard to baseline freezing until requirements and designs are fully developed. The rapid development method calls for the repeated application of configuration management with each system delivery. Further, a less strict application for the first deliveries and progressively more stringent application with later deliveries is called for.

▐▌ TABLE 5.16	Configuration Management—Identification
Definition:	Careful definition of baselines and baseline items for both hardware and software configuration items to be managed.
Question:	What is the current system configuration?
Answer:	The current system configuration consists of the following items: Item 1, Item 2 . . . Item N.

▐▌ TABLE 5.17	Configuration Management—Configuration Control
Definition:	The mechanism for preparing, evaluating, accepting, or rejecting proposed changes.
Question:	How are changes to the configuration controlled?
Answer:	The steps in processing changes to the configuration are: Step 1, Step 2 . . . Step N.

▐▌ TABLE 5.18	Configuration Management—Auditing
Definition:	The mechanism for determining the current state of the system configuration.
Question:	Does the system currently being developed meet the stated requirements?
Answer:	The system currently being developed differs from the stated requirements as follows: Difference 1, Difference 2 . . . Difference N.

Conventional configuration management consists of four major processes:

1. Identification
2. Configuration control
3. Auditing
4. Status accounting

Tables 5.16 through 5.19 summarize the definitions and basic questions at hand in dealing with these four components.

❚❘ **TABLE 5.19**   *Configuration Management—*
                              Status Accounting

Definition:	Tracking and reporting on all identified configuration items.
Question:	What changes have been made to the system?
Answer:	The configuration and changes to the system at this time are: Item 1, Change 1; Item 2, Change 2 . . . Item N, Change N.

Configuration Identification

The process of configuration identification involves the careful selection of all hardware and software items from the WBS over which configuration control is to be exercised. Maintaining control means that at any given point in time a report can be generated detailing what the current system consists of, what changes are being proposed, what changes have been accomplished, and what effects, if any, these changes have had on the ability of the system to meet original requirements.

The process also involves the definition of baselines appropriate to the development paradigm and the definition of the extent to which configuration management tools will be applied at each baseline. Baselines are established at discrete points throughout the implementation process to provide increasing degrees of control as the product development matures. As this maturation unfolds and further specific commitments are made, it is both possible and desirable to establish increased control over the emerging levels of detail.

There are at least five logical and discrete points at which to establish baselines for the purposes of configuration management. You may elect to use all of them or a subset depending upon the size, complexity, and length of the project and upon the level of confidence that all facets of the work can be adequately evaluated. They are the functional, allocated, design, product and operational baselines, as shown in Figure 5.23.

Functional Baseline— The functional baseline is typically established at the conclusion of the formal establishment of system functional requirements. The functional requirements document (FRD) is established prior to the product requirements review and in many cases provides the basis for a request for proposal (RFP).

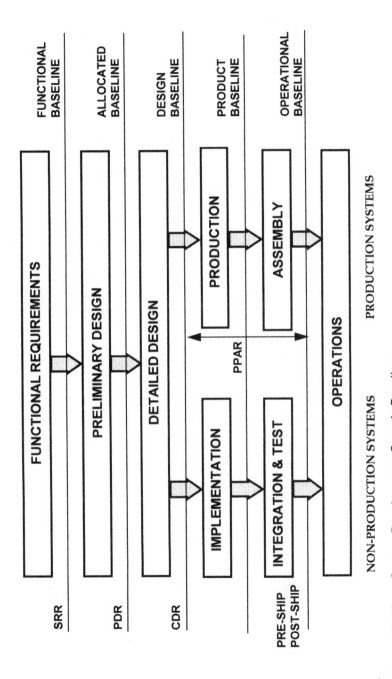

Figure 5.23. System Development Generic Baselines.

177

Allocated Baseline— The allocated baseline is normally established after the preliminary design review and prior to initiation of detailed design. It represents the point in the engineering process where required system performance has been allocated as system specifications to specific hardware and software configuration items and subsystems.

Design Baseline— The design baseline, when used, is established following detailed design and at the conclusion of the critical design review. This is the baseline to which all subsequent hardware and software implementation adheres.

Product Baseline— The product baseline is established following production. System and subsystem specifications, initial fabrication, system integration, and test and factory readiness are all complete.

Operational Baseline—The operational baseline is established as the system is delivered to the users and is the basis for all continuing system maintenance and product improvement phases.

Military programs typically use the functional, allocated, and product baselines. The functional baseline is established at the completion of the conceptual exploration phase. The allocated baseline is established after the validation phase and early in the full-scale development phase. In this approach, the product baseline hardware is often established prior to production, and product baseline software established at the completion of software code and test.

Configuration Control
As each baseline is established, control of changes to that baseline is also established. Formal methods of change control will now be discussed.

As previously noted, the items selected for control under configuration management are called configuration items (CIs). Logically, there are both hardware configuration items (HWCIs) and software configuration items (SWCIs). These are often defined at level four of the system at the hardware subsystem and software program levels as depicted in Table 5.3. This is a logical level on which to focus in non-military systems as well for identification of configuration items.

However, as always, the selection is not blindly driven. It is rather driven by the level of confidence in the degree of control required. Programmatic and/or technical management may wish to maintain close control over critical subassemblies or program components. Alternatively, control at the higher segment, or element, levels may also be appropriate.

The identification of configuration items is very important and worthy of considerable thought. The work breakdown structure, at all levels, can also be used as an effective guide in helping to consider and identify all items. The judgment of management here is key. Configuration items should be identified such that adequate control will result but the total number of items will still be manageable. Also, consider such aspects as level of functionality, testability, boundaries of discipline, and division of responsibility to physically distant organizations or contractors. Note that these considerations are similar to those evaluated in the allocation of work to specific subsystems. As always, the level of confidence with regard to maintaining visibility should assist in determining configuration item selections. It is not necessary to blindly apply the same level of control to all elements down to a particular level of the WBS. Control what you need to control. It is appropriate to use the concurrent design team as a focal point in the selection of configuration items and to review the selection with management.

Unfortunately, the whole business of configuration management and control is often considered a big burden. It is recognized that many organizations today routinely employ organization-wide configuration management strategies designed to support the staircase development paradigm. If the selection of baselines and configuration items is predetermined by organizational policy and does not fit a particular development paradigm in use, constructive efforts should be made to deviate or seek relief from the imposed policy during the Phase B planning stage. In Phase C, it will be too late. Changes should be proposed during this planning phase to both configuration management personnel and line management rather than to knowingly commit to definitions and procedures that may hamper the ability to perform during actual development. Remember, the more progressive product development paradigms require modifications to traditional staircase configuration management, including the timing of the tightness of control.

Another inconvenience often stated by product development teams is the common use of a standard change control board (CCB) that is organizationally external to the product development team. The problem arises because CCBs typically meet on given schedules. They also are concerned with other developments as well. This can cause delays in the approval of changes because CCBs may meet once a week, or even monthly. From the development team's standpoint, it would be highly desirable if the concurrent product development team were empowered to propose, evaluate, and implement necessary changes autonomously. A common argument against the PDT acting as its own CCB is a potential loss of objectivity—that is, a potential for conflict of interest that a third party would supposedly not exhibit. My position is that the PDT should be empowered to make any changes associated with implementation so long as programmatic issues, such as schedule and cost, are not impacted. This is important. Be absolutely sure to work this out during the planning phase, Phase B. If you don't, you run the risk of somebody, or even the entire development team, bypassing the system with the rationale of trying to get things done quickly.

Engineering Change Proposals Configuration control, in itself, encompasses the formal set of policies and procedures for requesting, evaluating, and accepting or rejecting proposed changes, corrections, and/or waivers to a configuration item. Figure 5.24 displays the salient steps in the generic process of change control. Requests for change can be driven by any unforeseen need—from necessity for design changes, response to changing funding profiles, to schedule changes and other engineering change proposals (ECPs). Requests for change can also be planned, such as installation of new operating systems, placement of new software deliveries on line, and implementation of preplanned product improvements.

Configuration control is maintained on the current baseline. The current baseline includes itself and previous baselines. ECPs may affect any or all of the functional, allocated, design, product, or operational baselines, depending on the time they are requested. Thus, changes that affect physical hardware and software also result in changes to all supporting documentation as well. That's why repetition in documentation should be avoided.

A single responsible entity is designated to log and track each ECP. This can be either the development team leader, the design team secretary, or a member of the design team. On large projects the responsibility may be assigned to a representative of the organizational CM structure.

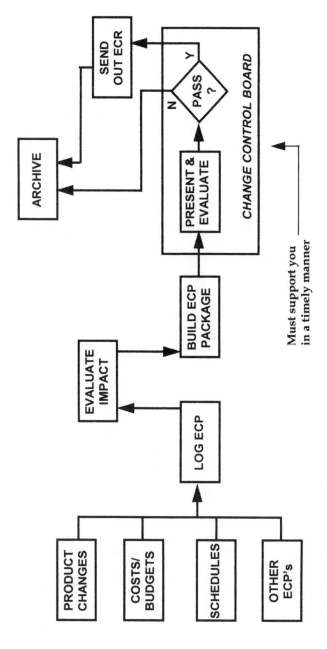

Figure 5.24. Engineering Change Proposal Process.

181

Assuming that the submission is a valid change request and not simply a failure report, or a misunderstanding, the development team then evaluates the impact of the ECP. The team usually restricts itself to evaluation with regard to technical impacts. Potential programmatic impacts, such as schedule and cost impacts, should also be brought to the attention of management. The development team then supports management in assessing the overall impacts.

The development team then prepares, or supports in the preparation of, an ECP package that will be presented to the formal change control board (CCB). CCBs usually have a number of members not directly connected with the project, which is both good and bad. It is good because, at least in theory, an element of objectivity is introduced into the deliberations. It is sometimes bad because the necessary understanding to make or break the case is not always present. While this may at times be a frustration, the condition is largely offset by requiring the proponent to make his or her case in a thorough manner. The best CCBs have representation from your CM organization, your project and programmatic management, all affected technical areas, a technically competent "outside" expert, and of course, the user or user representation.

The development team leader usually makes the presentation to the CCB, although a member of the concurrent team may also be the presenter. There should be appropriate cognizant engineer representation, as well as any other specialty, or technical backup, required to respond to detailed questions.

Whether the concurrent product development team acts as its own CCB for technical issues or the CCB is a separate organizational entity, all proposed changes must be separately documented. The content of a change proposal document, or presentation, should include

1. A statement of the problem or system upgrade proposed and a statement of the change requested. The latter should include the exact choice of language for the proposed change(s) to the affected specifications and/or requirements if appropriate.

2. A statement summarizing the analysis of all impacts including technical, programmatic, and any user impacts. Don't forget the later; it is the most important.

3. A tabulation on separate slides of the advantages and disadvantages of ECP implementation versus alternatives, even if an entry under one of the headings is "none."

4. A recommendation to accept or reject the ECP, supported by a succinct statement of consequences.

Lengthy presentations should be avoided. A basic presentation may consist of four to six slides, with up to twenty backup slides.

CCB meeting minutes are also recorded by a designated CCB secretary. The minutes should include a list of all in attendance, their affiliations and phone numbers, a brief but complete account of proceedings, and any technical or programmatic supporting material necessary to provide a self-contained representation of what transpired.

Following the CCB meeting, an engineering change request, ECR, is issued by the board if the proposed change has been accepted. The ECR constitutes official authorization to generate detailed change implementation plans and to initiate the change. The progress of change implementation is tracked by a single designated responsible person. All ECPs and ECRs are archived by configuration management.

Accepted changes are often classed with respect to their impacts and priorities regarding importance and need. The priority largely determines the time frame in which the change will be implemented. Two common priority schemes for the handling of ECPs and ECRs, as well as FRPs, are listed in Table 5.20. The classifications generally effect the importance and hence the timing of the change to be made.

Failure Reports Failure reports (FRPs) are also logically handled by configuration management for at least two reasons. First, a record of outstanding FRPs and closed FRPs is a widely used metric for partial reporting on the status of an implementation effort. Secondly, it is not uncommon for FRPs to be written that request features beyond the stated requirements—that is, requests that actually should have been change proposals.

TABLE 5.20 *Common Priorities for Handling Change*

Two Level
1. A change is required in form, fit, or function.
2. Changes are required in document language or in minor component substitutions.

Four Level
1. Prevents mission accomplishment.
2. Seriously degrades mission accomplishment.
3. Minor impact on mission accomplishment.
4. Documentation change only.

In principle, FRPs can be submitted by anyone. However, the point in time at which a "failure" occurs is usually considered to be *after* formal delivery of hardware or software. The term "delivery" refers to a specific point in time at which developers consider that the product is complete and hand it over to another party, such as an independent testing group or end users for acceptance testing. That is, programmers in one development group typically don't generate FRPs upon themselves. Their interactions are generally less formal (as they should be) to allow responsiveness in identification and correction of "bugs."

In more rigorous settings, I have seen FRPs being generated during development by a programmer in one organization against a programmer in another organization. This happens when large programming staffs are involved across organizational entities as specific programs are provided for cross support. This condition is one of the things that the concurrent product development team is designed to avoid. When programmers interact more intimately, however, there is a fine line between simple debugging in unit test and identification of a "failure." A failure is quite naturally not perceived as a failure during early routine debugging activities.

An alternative approach to record keeping during development, purposely designed to be less stringent, involves the use of anomaly reports (ARs). ARs are used primarily to maintain a log of problems encountered so that the experience can be put to good use in the future. This is a good idea, in theory, but is difficult to effectively implement. Programmers don't like to fill out forms. Still, it is often useful to maintain records of significant unforeseen difficulties that may help you and your programmers in future similar efforts. The use of ARs is one method of doing this. Another method, which we all should have learned in school, is to have your programmers maintain the classic engineering notebook. This is a good habit and is particularly valuable when dealing with new designs—that is, designs that involve the organization of knowledge, algorithms, or data structures not encountered before.

A standard method for the handling of failure reports is depicted in Figure 5.25. FRPs, are routinely submitted to the concurrent product development team first, where they are logged. On small projects, this may be done by the team leader. On larger projects, designate the design team secretary to log and track all FRPs.

Each FRP should address only a single issue. An FRP form that addresses multiple problems by listing more than one at a time should be resubmitted with a single form for each perceived failure. The reason for this is that as each problem is successfully addressed, the identified FRP can be closed and partial closures of FRPs can be avoided.

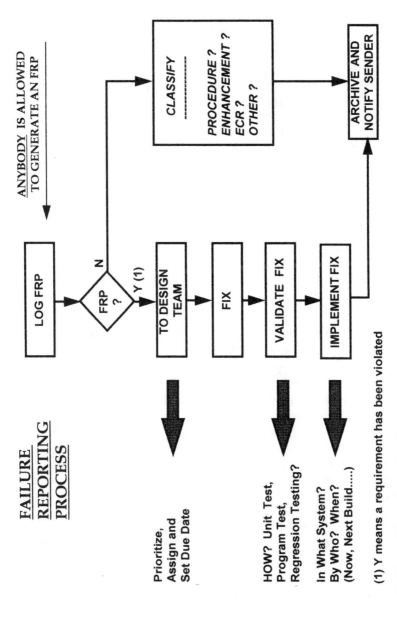

FAILURE REPORTING PROCESS

ANYBODY IS ALLOWED TO GENERATE AN FRP

LOG FRP

FRP ?

N

Y (1)

CLASSIFY
_____
PROCEDURE ?
ENHANCEMENT ?
ECR ?
OTHER ?

TO DESIGN TEAM
Prioritize, Assign and Set Due Date

FIX
HOW? Unit Test, Program Test, Regression Testing?

VALIDATE FIX

IMPLEMENT FIX
In What System? By Who? When? (Now, Next Build.....)

ARCHIVE AND NOTIFY SENDER

(1) Y means a requirement has been violated

Figure 5.25. Failure Reporting Process.

An initial assessment of the FRP is made to determine

1. whether the report records an actual failure,
2. whether it requests a change to existing requirements, or
3. whether the FRP was generated by a misunderstanding of how the product operates or should be used.

By definition, a failure is specifically a deficiency in meeting a single measurable documented functional requirement, or specification derived therefrom. Requests for changes in existing requirements should be handled through an engineering change request (ECR). Misunderstandings should be handled through appropriate notifications, and/or change requests for help files or user documentation. Failure reports are valid only when reporting actual failures to meet established requirements.

The concurrent development team leader is responsible for determining acceptability of an FRP. The status of each submitted request should be routinely reviewed with the concurrent product development team. If the submission is not a valid FRP for any of the reasons given above, the reason is noted on the appropriate section of the FRP form and the originator receives a copy. The generator may then take appropriate action in response. All FRPs, regardless of their content, are archived.

If the FRP calls out a single bonafide failure, a plan of action is then agreed upon by the development team. This action consists of setting a priority for correction, designation of an assigned party responsible for correction, and the setting of a correction due date.

From this point on, there is only one copy of the FRP that is official. The development team leader, or a designee, acts as sole custodian for the official version. Any duplicates of the official copy that are made in the interim prior to FRP closure are not official. Copies are distributed for information purposes only with regard to the latest status of progress toward closure. Entries in fields of the FRP form made on distributed copies have no validity. Official entries of information in any field of an FRP form can only be made by the team leader, or a designated custodian, and only on the original copy. If a separate organizational CM entity is tracking your FRPs, you must work out an agreement for a strategy of control for a single official FRP.

On a given project, priorities for the correction of valid FRPs are generally the same as those adapted for the handling of CRs shown in Table 5.20. The responsibility for correction is generally assigned to the cognizant engineer for the area affected. As with any action item, this does not mean that the assignee will necessarily make the correction firsthand. He or she may assign the accomplishment of the correction to someone else. But the principal assignee is the sole responsible party for the realization of the correction and testing of its completeness at the unit level on or before the agreed-to due date. The person making the correction is not responsible, unless that person is in fact the assignee. Only one person is responsible.

The setting of the due date is primarily driven by the needs of the user, or other developers, and is negotiated, as are priorities and correction responsibilities at the concurrent product development team meetings. Potential schedule impacts on other concurrent work need to be identified and discussed. The guiding concept in all of these determinations should be to accommodate the user, or team member, as much as possible. But also be very sure that the assignee is comfortable with the due date. Unrealistic assignment of a due date should be avoided. Bean counters love late FRPs; they absolutely love them. And they will find you.

While the fix is being carried out, a plan for acceptance testing of the fix and its ultimate implementation on the system must be specified. These plans can be quite simple or they can be genuinely complicated. On projects of any size, testing and implementation schedules are usually determined by previously scheduled build deliveries.

Realistically, cases sometimes arise when even our best priority schemes become replaced with "do it immediately." Deviation from accepted practices and our best-laid plans suddenly becomes the order. In the "quick fix" setting, the trick is to determine whether the fix is related to a small and isolated hardware or software condition or whether the fix has systemwide effects. The great danger is assuming the former when the later is the case. It is more difficult to make this determination in software products than in hardware products. The safe approach, as always, is to do a lot of testing at the development level—right at home—before implementation in the operations environment. If a quick fix is truly required, assemble the concurrent product development team together for a special meeting. Discuss all aspects of the problem. Coordinate the special actions you must take by making specific assignments and devise contingency support plans should the issue persist. Use all the time that is available.

Liens Liens consist of promised hardware items or software capabilities that remain outstanding at the time of a delivery. They are usually fairly minor items for which workarounds have been devised. Liens are accepted (tolerated) by those who receive hardware or software deliveries because it may be more desirable to proceed with scheduled testing or operations minus the full complement of system capabilities than to wait for all promised items to be provided. Thus, the decision to execute a delivery that is deficient in one or more items is primarily made by the entity that is to receive the delivery.

Liens arise for a number of reasons. Procurements may be late due to vendors slipping the dates of deliveries of your project, financial decisions out of your control, or any number of procurement problems. If previous deliveries have already been made, there may be outstanding failure reports that have not been corrected at the time of delivery. Liens, in one form or another, against documentation errors or deficiencies are also common. Action Items generated at any of the reviews held prior to the first delivery or between deliveries may not have been adequately resolved. Finally, unforeseen technical difficulties can easily cause schedule changes that impact the ability to deliver all required items on schedule.

Note that liens are written against requirements. They are a part of an agreed-upon strategy to forgo the meeting of a requirement at the time of a delivery. Thus, change requests cannot become liens unless the change has formally been incorporated by the configuration management change control board as a new requirement to be met by the project, and a delivery date incorporating that change is to be scheduled for a later date.

Action to remove liens is negotiated by the concurrent product development team in the same manner as failure reports. Each lien is assigned to a responsible party for its resolution. A priority is assigned and a due date is established. For consistency, it is expedient to use the same priority scheme that you use for failure reports, change requests, and action items.

The listing of liens is maintained by the systems engineer. Figure 5.26 provides an example of a lien list.

The sample lien list shows outstanding liens as of 6/4/200_. As each lien is closed, its closure is noted in the next reporting period and, then, it is subsequently dropped from the list to avoid carrying excessive outdated information. The complete history of lien handling, however, is maintained in a computerized master lien list database. The reporting periods are generally one week apart, as the lien list is routinely reviewed at each team meeting.

An identification number for each lien is given in the first column. The second lien in the list, lien number 5, is seen to have been closed during the week prior to 6/4/0_. (See note in comments column.) This lien will not appear in the next report. Also implicit in the

Page 1 of 1 Report Date: June 4, 200_

No.	H - Hardware S - Software	Assign Date	Lien	Assigned To	Due Date	Comments
1	S	2/4/9_	PACK routine	J. Smith	7/12/0_	
5	S	2/12/9_	COPY routine	R. Jones	6/19/0_	Closed 6/1/0_
7	S	3/18/9_	Tape to tape routine	M. Gorden	8/13/0_	
8	H	5/2/9_	Terminal	L. Scott	6/1/0_	Vendor delay, six of seven delivered, seventh due 7/15/0_
9	S	5/2/9_	Users Guide Doçument	J. Smith	7/1/0_	Clarify Sec. 7.1, See AI #23
10	H	5/15/9_	Range Bin # 4	L. Scott	7/15/0_	In repair

Figure 5.26. Lien List Example (use also for FRPs, action items, etc.).

list is the fact that liens 2 through 4 and lien number 6 were closed in earlier reporting periods and these records were dropped from the database report form.

Lien number 8 is currently past due. The reason for this condition is summarized in the comments column and a new anticipated closure date is shown. Lien number 9 calls for a documentation section rewrite for clarification, which resulted from a review action item (AI #23), and lien number 10 resulted from an unforeseen hardware failure.

Configuration Auditing

Configuration auditing is the process of determining the extent to which the current configuration baseline conforms to the previous baseline and to the original functional requirements (functional baseline). The process involves both verification and validation, so-called V and V. These terms are widely used but often interchanged and more often misunderstood. Verification is the process of determining if the configuration items of a given baseline meet the requirements of the previous baseline. This is needed. But note that simply verifying that the more developed configuration items of a given baseline are traceable to a previous more generic baseline is not a guarantee. Verification, in and of itself, does not establish whether baselines have strayed through time from the original intent of the system. That's where validation comes in. Validation is the process of determining if a given baseline still solves the intended problem. Validation is concerned with traceability to original mission objectives. Putting it more simply, does it work right?

Both verification and validation are best tracked through the use of traceability tables. For the purposes of verification, the tables trace forward and backward between baselines. The validation tables trace each baseline, as it is established, both forward and backward to the original functional baseline. Two sample tables to assist in validation follow. Table 5.21 shows a sample portion of one traceability table that consists of a succinct summary of functional requirements. This table is constructed directly from the functional requirements document and serves as an easily handled quick reference for each requirement. The functional requirements document (FRD) itself, of course, contains each requirement in it's entirety and can be referenced as needed for clarity. The requirements summary table provides a terse description of a single requirement for each appropriate paragraph in the FRD.

▐ TABLE 5.21	Sample Portion of Functional Requirements Summary

FRD Paragraph	Requirement
1.0 ___	
2.0 ___	
3.0 ___	
3.1	Applications Processing Requirements
3.1.1	Generation of Plots
3.1.2	Data Conversion from ASCII to EBDIC
3.1.3	Contrast Manipulation
3.1.4	Two-Dimensional Fourier Transforms
3.1.5	Map Projections
3.1.6	Photographic Hardcopy
3.1.7	___
3.1.8	___
4.0 ___	
5.0 ___	

Table 5.22 is a sample portion of a traceability table for each baseline. This table records the specific action taken at a given baseline in response to each original requirement by paragraph. The two tables accompany each other and are used together. For example, in Table 5.22 we see an example of traceability of each FR paragraph to a paragraph in the software specification document (SSD) as well as the hardware items and software modules intended to meet the requirement.

Thus, the software program called QPLOT1 is shown to respond to the functional requirement stated in the FRD in paragraph 3.1.1. Table 5.21 shows that paragraph 3.1.1 of the FRD covers a requirement for the generation of plots. Table 5.22 indicates that the specification for QPLOT1 can be found in paragraph 4.2.3 of the SSD. Note that the SSD paragraph numbers are not necessarily in sequence as are the FRD paragraph numbers. This highlights the fact that the traceability carried out in this example is forward from the FRD to the design baseline.

The right-hand three columns of the sample traceability Table 5.22 indicate the specific hardware items and software modules that respond to each requirement and also provide an area for comments.

▌⁞ **TABLE 5.22** *Sample Portion of Design Baseline Traceability*

FRD Paragraph	SSD Paragraph	Hardware	Software	Comments
1.0 ___				
2.0 ___				
3.0 ___				
3.1				
3.1.1	4.2.3		QPLOT1	
3.1.2	5.1.1		AECONV	
3.1.3	5.2.4		CONMAN	
3.1.4	5.2.5		FFT101	
3.1.5	6.3.1		MAPPR2	
3.1.6	4.3.3	FOTO	FOTOHC	Two FOTO units to be used.
3.1.7	___			
3.1.8	___			
4.0 ___				
5.0 ___				

Backward traceability consists of making sure that all baselined hardware and software items have been accounted for by forward tracing and that no items exist for which there is no documented need. This exercise can be carried out using traceability tables of similar construction but designed to trace in the opposite direction. Both forward and backward tracing should be accomplished in the interest of completeness.

Traceability tables are intended to assist in organizing one's thinking in carrying out the auditing process. The mechanical use of these tables does not, in itself, constitute auditing. Careful thought regarding the adequacy of elements with an increasing level of detail in successive baselines is the real function of auditing. Successive levels of detail should be clear and unambiguous so that a lucid forward and backward mapping between baselined items and to and from functional requirements is evident.

Effective auditing ensures

1. The unfolding design and implementation is responsive to actual needs.
2. Unspecified capabilities are not finding their way into the system.

❚⁞ TABLE 5.23   *Content of the Status Accounting Database*

1. A description of C1.

2. The location, identification number if appropriate, and cognizant engineer for each CI.

3. A listing of each CI in each baseline and the date that each baseline came into being.

4. ECRs and their status of the reporting date. ECR status includes approved, disapproved, progress of approved change implementation, responsible party, priority, and due date.

5. FRs and their status as of the reporting date. FR status includes responsible party, priority and due date or closure date. These records also provide a base for tabulation of FR generation and closure rates.

6. Action items generated as a result of audits or formal reviews and their status.

7. Any other information you may deem advisable as a result of configuration identification, configuration management or auditing functions.

Configuration Status Accounting/Reporting

Programmatic and technical management must be capable at any time, on demand, of easily updating configuration status and generating a succinct report for its own use or for the use of management.

Status accounting is an administrative function that maintains a formal record of each of your configuration items including their definitions, change requests, change approvals and disapprovals, progress toward implementation of approved changes, implementation completions, and actions taken as a result of configuration audits.

The status accounting mechanism should be computer based to support all members of the development team. A computer-based system also allows all personnel to stay current in a convenient manner. The status accounting database should include the data shown in Table 5.23 as a minimum.

Configuration Management of Subcontractors and Vendors

Large vendors usually have their own configuration management organizations and practices. Small ones may not. When issuing a request for proposal, or a request for quotation, it is advisable to include a separate section asking prospective vendors to state how they intend to meet your project's configuration management requirements. Generic configuration management requirements should be stated clearly but not in such detail as to require major modifications in the way the vendors currently do business. Evaluate vendor responses numerically as a separate part of your vendor acceptance criteria algorithm. If all aspects of a vendor's response are satisfactory except its approach to configuration management, it is always possible to impose your own system. It is wise, however, to avoid this since it is only likely to increase the cost. I am reminded of one parent company that was the major contractor for a product development effort for a prominent national sponsor. Substantial procurements were involved. The parent company's request for proposal (RFP) included configuration management system requirements that copied its own system verbatim. The procurement was supposed to cost around $40 million. Four responses came in—all over $50 million. One of them was $65 million. Why? They all had to build brand new complicated CM systems. The development was canceled. The parent company was dropped. Two years were wasted. The sponsor went someplace else.

Good organizations know how to do configuration management in their fields. Contractors and vendors should be given an opportunity to specify how they intend to meet the project's stated configuration management needs. If required, negotiate with the idea of maintaining their basic approach and procedures. Finally, include in the selection criteria a means of eliminating those who fail to exhibit an ability to maintain effective configuration management.

The Configuration Management Plan

The configuration management plan (CMP) formally defines and describes the policies and procedures to be employed in the execution of CM. The CMP also defines organizational structures and responsibilities for CM functions. This section provides a guideline for the construction of the CMP.

The CMP may be imposed on a given development effort by an external CM structure by directive or by the product development team itself. If the former is the case, the plan should be reviewed with great care. CMPs written to apply to any and all projects can be designed to cover all the bases in such detail as to result in an overbearing regimen. Still, such documents are valuable as guides. Should the development team leader feel that specific modifications or waivers are in order and if the leader can present sound reasons for their adaptation, it is not unreasonable to request such actions. Good CM plans balance the need on a project-by-project basis for individual (or team) creativity against the need for increasing levels of control through time.

The more creative phases of a project occur earlier in the development cycle. The need for creativity diminishes as the design matures. The most freedom is required between the development of functional requirements and the preliminary design. While strict control is maintained over the functional requirements document, the freedom to determine issues of "how" in response to the "what" of the functional requirements should ideally be maximized within the functional requirements envelope. In the staircase development model, the work accomplished between the preliminary design and the detailed design is naturally restrained to a more limited envelope. When there are fewer options, the need for control becomes increased. Beyond the detailed design, after passage of the critical design review, control typically needs to become rather stringent. This is a simple top-level concept and should be a consistent element in guiding the approach to CM implementation.

Table 5.24 provides a suggested outline for a generic configuration management plan. Comments on each section of the outline with suggested inclusions follow:

Introduction—Provides a one-sentence statement of the document's purpose. Include a statement of scope, if appropriate, and include a subsection covering applicable documents.

Organization/Responsibilities—Show the organizational relationship of CM to project management and show an organizational chart

---

▯⁞ TABLE 5.24   *Configuration Management Plan Outline*

1. Introduction
2. Organization/Responsibilities
3. Configuration Identification
4. Configuration Control
5. Configuration Status Accounting
6. Configuration Auditing
7. Subcontractor/Vendor Control
8. Glossary of Terms

of the CM function. Include the change control board's organizational relationship. Discuss responsibilities of management, the CM organization, and your design team.

Configuration Identification—Define the baselines to be used in the project. Clearly define the beginning and end of each baseline in terms of deliverables, completion of documentation, and/or completion of reviews. Explain the philosophy to be used in designating hardware, software, and documentation as configuration items. (The actual designation awaits the execution of the plan.) Specify the plan for use of identification numbers for each level of configuration items.

Configuration Control—Explain the plan for the systematic evaluation, coordination, approval/disapproval, and subsequent handling of proposed changes to any baseline. Define the entire change review and implementation process as discussed above. Tailor, or expand this, to your need with a supporting block flow diagram. Cover all functions of the change control board including the CCB review of deliveries and placement of hardware and software on-line. Provide sample forms, in an appendix, for each form you intend to use such as failure reports, requests for change, anomaly reports, documentation change notices, implementation notices, and the like. Include directions on the reverse side of each form as to how to fill them out, the purpose of the form, and the organizational entity to whom they should be submitted.

Configuration Status Accounting—Specify the form and format for each item to be covered. Specify all database characteristics such as sorting capability, report generation capabilities, and the like, that you may require in support of your ability to conduct full or partial audits.

Configuration Auditing—Define all schedulable formal audits. As a minimum, a formal audit should be scheduled in support of the acceptance of each baseline. Specify the need for unscheduled partial audits that you may wish to conduct on demand, such as statistics on FRP closures, changes in progress, changes implemented, and so forth.

Subcontractor/Vendor Control—Specify your plan to control subcontractors and vendors. If you are using their plans, provide a summary description and refer to their documentation. Describe modifications to your own control procedures that are acceptable and have been agreed upon. Describe all interfaces and contacts between your CM status accounting process and theirs. Describe all auditing practices, scheduled and unscheduled, that you intend to carry out on vendors and subcontractors.

Glossary—There's a lot of jargon in CM. Include a glossary of terms.

PLANNING FOR LOGISTICS SUPPORT

Logistics support addresses all issues associated with the execution of support activities required for both the development and operation of the mission product. Logistics support consists of designs and procedures for the following items:

- Product maintenance plan
- Supply support
- Test equipment
- Transportation and handling
- Technical data packages
- Facilities
- Personnel and training

Keep these seven elements in mind for a moment. They will be discussed in detail shortly. But, first, a few general comments on logistics support.

The development team leader, and the team, must consider two distinct logistic support strategies—one for the sole support of development activities and another to support field operations. The former is generated to support the design and implementation process itself. The latter is a part of the product requirements. They are documented in the functional requirements document and flow down in more detail as part of the product design for support of the product after delivery to the user(s).

Operational support strategies can have a profound impact on system design and on system cost tradeoffs. Systems that, upon failure, are difficult to troubleshoot and restore may require highly trained personnel and test facilities in order to maintain availability requirements. These designs increase operational costs. Alternatively, operational support concepts based on self-diagnosing modules that are easily replaced can significantly alleviate requirements for training, special test equipment, and complex spare inventories in the operational setting. But this approach adds to development costs.

Logistics support concepts are first defined prior to the product requirements review. The development of these initial concepts involves major decisions that have direct impacts on the overall product design. As with all design decisions, assistance in the making of major logistics support decisions can be effectively guided through consistent review of the prioritized competing design characteristics for the system under analysis. The logistics support analysis progresses to a preliminary logistics support plan prior to the preliminary design review, and finally matures to a complete plan that is prepared for detailed examination at the critical design review.

Now let's return to our discussion of the seven logistics support elements.

Product, or System, Maintenance Plan

Reliability and maintainability analyses are conducted to support the system maintenance plan (SMP) and to ensure that system availability requirements will be met for both development and for operations. Life cycle cost considerations are an important part of the design of the operations SMP. Over the system lifetime, total operational costs can be greater than development costs.

The issues can rapidly become complex. It is a peculiarity that in some large systems the fundings for development and for operations come from different costing centers within the same procuring agency. In such instances, motivation to minimize development costs can actually take precedence over life cycle cost considerations such that higher total lifetime costs result. There are other instances in which the developer does not directly assume maintenance costs. In the development of commercial products (cars, appliances, locomotives, etc.) where the consumer typically assumes maintenance responsibility, it is wise to consider maintainability as a selling point. Still, in the marketing of less expensive products with a relatively short life (calculators, fashion watches, etc.), a valid approach may be that maintenance is replaced by simple unit exchange. In some cases, manufacturers have found it cost-effective to take this one step further by dropping their quality control function all together. A higher ratio of defective units is sent to market and the consumer, in effect, carries out quality control by returning defective units for exchange. This approach, however, is not in keeping with modern quality management trends that place a premium on customer satisfaction. (As consumers, none of us appreciate this approach.)

In most systems of any reasonable size, it is almost always expedient to implement designs that reduce operational costs when possible. One common approach to the reduction of operational costs is in the use of built-in test equipment (BITE). BITE consists of special test equipment that is built into a functional unit and is usually executed by special test software. The addition of BITE capability, while increasing development costs, is typically employed to reduce maintenance costs. The use of built-in test (BIT) is an effective way to reduce the time for fault isolation and hence the requirements for training in forward service areas. Because it can reduce the mean time to restore (MTTR), it can also provide relief on requirements for module reliability.

The design of BITE and BIT involves the modular organization of hardware into units that can quickly be diagnosed and replaced. These units are often referred to as line replaceable units (LRUs) because they are replaced at the "front" line of the operational setting. In this support scheme, faulty LRUs are removed and shipped to an intermediate support area where repair and recalibration can take place. In the military, these intermediate areas are often highly mobile and limited. Units that cannot be repaired at this level are returned to the depot level, where complete repair capabilities are available.

The analysis that supports the design of such three-tiered systems is not so concerned with the fate of a single LRU as it is with the maintenance of inventory flow up and down the system. Required inventories at each of the forward, intermediate, and depot levels are determined by LRU and system availability calculations and/or experience.

Operational maintenance planning involves the definition of maintenance concepts to support the mission product in operations. The maintenance plan must weigh the reliability of the design with the timeliness of the support mechanism. That is, it should be based on a thorough reliability/availability/maintainability trade-off analysis.

System development maintenance planning similarly entails consideration of each of the elements of logistics support such that the design and implementation process is efficiently supported. Maintenance planning, of course, draws most heavily on reliability, availability and maintainability (RAM) analysis.

In less stringent government and commercial systems, the support mechanism will characteristically employ only two levels, consisting of on-site support and external vendor support. In many large systems, vendor support is permanently on-site in the form of customer engineers (CEs). Such CEs have permanent offices at a preferred customer's base of operations and use this base to service the main customer as well as other customers in the immediate area. If you are involved in a system that could fit this model, it is worth looking into the vendors' policy. When the CE is at the site, the logistics factors of repair and/or replacement are consistently reduced in support of operational availability.

The simplest maintenance scenario entails vendor support on a call-as-needed basis, usually supported by a maintenance contract. When system availability is not critical, this can be a satisfactory arrangement. However, you need not restrict yourself to a single strategy for all time. For example, a ground support system for a spacecraft during planetary encounter may employ a combination of increased spare unit inventories along with intense on-site vendor support during this critical period and, then, revert to a less demanding support mechanism during less critical cruise operations. The same is true for city fire department dispatch centers on the Fourth of July, typically their busiest day of the year. The effects on system availability of such strategies can be estimated through the attending modification of repair and/or replacement times for each of the units affected.

Vendors, of course, do not commonly allow users to attempt to repair their equipment and often stipulate this condition in maintenance contracts. In these cases, mean down time (MDT) can be significantly reduced by maintaining spares on-site, such as spare terminals, cables, keyboards, printers, displays, and the like. On-site personnel belonging to the user organization can swap such units in and out without violating contractual arrangements. Mainframes are usually too expensive to maintain idle backup spares. Many systems, however, employ one mainframe for operations and a second for software maintenance and continued development. With this arrangement, the development machine can serve as an operational backup. This arrangement greatly reduces product MDT, hence increasing system availability.

It is evident that the designer of the reliability/availability/maintainability (RAM) component of the SMP has many options. The basic task is to carry out a balanced RAM analysis in the context of the complete life cycle cost setting. Admittedly, no small task. As with all such analyses, however, the work should first be carried out at a top level in order to gain overall insight into potential problem areas. The analysis is then extended in detail as required and as the design matures.

The starting point is to realistically set the system availability functional requirement. This requirement must be determined by the product user within the concurrent product development team setting. The next step is to perform a RAM analysis on the functional system architecture upon which the design team has agreed in order to meet functional performance requirements. The functional architecture does not have to be absolutely finalized. It may change as the design process proceeds, but it is better to start as early as possible. On the other hand, it is not usually productive to start your RAM analysis on some arbitrary architecture that is not driven by an availability requirement.

If meeting the product, or system, availability requirement on the performance-oriented architecture is difficult (it may not be), then mean time to restore (MTTR) and/or mean down time (MDT) issues need to be addressed. Work on the weakest link should be done first, using combinations of higher reliability parts, redundancy, and/or MDT reduction concepts until the next weakest link becomes evident. Then work on that one. Work in this balanced fashion until the system availability requirement is met.

The analyst must also make top-level cost estimates for each alternative considered. Employment of higher-reliability units, redundant units, and strategies to reduce MDT all cost money. Both development and operational costs for each proposed alternative need to be estimated. BIT, for instance, increases development costs significantly but can reduce operational costs to the extent that life cycle costs are actually lower. As a guide to considering MDT costs, review the other basic elements of logistics analysis, which include supply support, required test equipment, transportation and handling, technical documentation, facilities, personnel, and training.

When a RAM strategy for meeting the product availability requirement has been determined and attending costs have been estimated, there may be a conflict between perceived cost guidelines and product availability requirements. The team leader then turns to defined product priorities to determine the best course of action. The previously agreed-upon priorities should provide a definite guideline as to which characteristic must be subordinate (within limits).

Supply Support

In the development domain, support equipment refers to those equipment items—including tools, spare units, spare parts, inventory levels, and consumable items—required for system development and maintenance of development equipment, or portions of the product, under development.

Operational support equipment categories are similar, but are identified and/or designed as a part of the development effort to provide eventual operational support for the fielded system. The locations, types, and quantities of support equipment will be different for the development and operational environments. In either case, they are defined by the logistics analysis effort.

Test Equipment

Test equipment in the development setting includes electronic test equipment, instrumentation, diagnostics and diagnostic equipment, mockups and special test rigs for fatigue, and environmental and performance testing. This includes unit testing, integration testing, and product-level testing during development. This also includes all contractor support and consumables associated with all phases of development testing.

In the operational setting, test equipment includes electronic, diagnostic, and other test equipment to be used in the field by maintenance personnel. If a multitiered operational support philosophy is used, test equipment applicable to each level is required.

Transportation and Handling

Transportation and handling (T&H) in logistics support is no small matter. For example, an informative incident related to the failure to adequately consider the impact of transportation and handling on system design deals with a system developed for the shuttle cargo bay, implemented in California by experienced engineers. The system wound up being delivered to Cape Canaveral by rail and barge when it was discovered too late that it wouldn't fit through a C5-A door.

Another incident of interest involved a group that made arrangements for delivery of a number of large antenna towers in Texas for propagation testing of a prototype communications product during development. The engineering details were extremely well planned, but when the crew got to the gate with three big trucks, the security guard had never heard of them. The guard made a few calls, but the contact point was away at that time. By the time security determined who the strange visitors were, the remainder of the day and half of the following day had passed—which the crew spent happily around the pool at the local Holiday Inn. Who was the systems engineer on that one? Me. It hit me that day waiting around the pool in Texas that logistics support, in it's entirety, is like a smart animal. It lives out there, pacing back and forth, waiting for you downwind. If you don't stalk it—it will find you.

Transportation and handling has to do with every aspect of equipment storage, packaging, and preservation, and the total logistics of moving it from one place to another. The impact of T&H on design can be anywhere from minimal to highly significant. But there is almost always some impact, whether the product be consumer oriented or a high-tech military system.

It is quite common for acceptance testing to take place after installation. This means, by definition, that your entire system will have to be picked up and moved before the customer signs off. This operation is still technically in the development phase. It is not unusual for projects to suddenly consider these aspects of design deep into the design phase—dangerously late. On-site prototyping and feasibility testing of subsystems as well as customer training exercises may also be a part of the development plan that may require further transportation and handling considerations.

When operational mobility is a system requirement, transportation and handling will clearly impact a design. Here the concern is with the ability to rapidly pick up, package, and move the product through all vibrational and environmental conditions to be encountered and to reestablish operations at a different location. Early and thorough determination of design impacts on size, weight, mountings, and so forth, can avoid expensive workarounds later. The placement of hooks at centers of gravity, or other strategic locations, as well as accesses for fork lifts, handles, wheels, and slides can all be unfortunate appendages when designed as afterthoughts. Even when operational mobility is not a requirement, equipment still has to be designed to ship, be returned, and be shipped again without harm.

There are also cost implications related to the efficiency of transporting supplies and materials in the operational setting. The optimization of routing and the number of trips required are but two considerations in the design of a logistics support system.

Technical Data Packages

Technical data packages are a critical element of logistics support. These packages coordinate technical documentation for all items at all levels of the work breakdown structure.

During the development phase, technical data packages coordinate the technical documentation for all user needs, requirements, specifications, interface, management plans, options analysis, special study, development logistics support, configuration management, testing, hardware and software design documentation, and anything else that supports development. The majority of development technical data packages is never delivered to the customer.

Operational technical data packages are deliverables. They typically consist of instructions for assembly, installation, troubleshooting, operations manuals, user manuals, maintenance manuals, training manuals, and transportation strategies. In constructing operational support documentation it is important to make a clear distinction as to the aim. One aim is to provide a reference document. Reference documents are excellent for people who already understand the basic product mission and a fair amount of detail regarding hardware, software, and operations. *But reference documents are of absolutely no value as learning aids.* (That's why we all feel so stupid when we read typical commercial software documentation.) They serve principally to remind the user of items forgotten, or to clarify procedures the user is already familiar with. (Dictionaries don't teach English, nor are they intended to.)

Another distinct aim of documentation is to instruct. Teaching documents are totally different from reference documents. They take the reader through a clear, top-down, descriptive process that only gradually comes to significant detail. Teaching documents are typically very difficult for engineers to write. When an author knows a lot about something, it is very easy to take fundamentals for granted and wind up instructing at a level inappropriate to the student audience. Software documentation is notorious for this. Most engineers have not had formal teaching experience, nor are they motivated in these directions. Good technical people, by nature, want to get on with their own endeavors and to keep pace with their own technical and professional issues. Stopping the fun work in order to generate instructions so that people who may be months, or even years, behind them may understand basics is an incredible imposition on them.

Technical writers, on the other hand, accept this kind of assignment by trade. A good one is well worth the time and money to meet schedules for the quality production of training and learning material. A common argument against the use of technical writers is that they "don't know the product" or "the learning curve will take time." The fact is that unfamiliarity is a distinct advantage in that it forces treatment of the subject from square one—exactly what is desired in a training (learning) document. Good technical writers have broad technical backgrounds and writing skills in which engineers are notoriously weak. (They can do things such as spell and construct English sentences.) Technical writers can be particularly useful in the generation of documents designed to instruct.

A third distinct type of documentation is design documentation. In some cases, design documentation, or portions of it, is used to support technical manuals delivered to operations. Problems in generating this kind of documentation for mechanical and electronic hardware items are generally not too severe because designers need layouts, drawings, string lists, logic equations, topological illustrations, or schematic representations in one form or another to do the job. It is a natural part of the construction of hardware.

This is not true in programming. A programmer can go to a terminal and simply start coding without any requirements or design documentation—and you can count on their doing exactly that. They "design" by "coding" because this is the most fun.

A few years ago, I had one programmer threaten to quit rather than generate his own design documentation; forget reference or training manuals. Now I realize that the problem was of my own doing. I was asking someone to do something that he had no professional interest in doing. Nor did he have the capability. No one likes to do something he or she doesn't know how to do. Further, I had no direct control over his paycheck in the matrix organization, and the line management would not consider replacing him. His prodigious ability to produce working code for perplexing problems was unparalleled. He was simply perceived as being too valuable. "Do it or else" was not a viable solution. The result? An unforeseen cost was added to my budget when management (at my recommendation) chose to hire a technical writer to get the job done. Now I think about these things up front.

All technical managers must address the problem of software design documentation. In the end, the only way programmers can be motivated to document requirements and design as they go is to employ a methodology that it is accepted by all as an advantageous way to proceed. If there is no clear advantage to integrating software documentation into the design process, programmers will always tend to document last, and then only with considerable nudging and cajoling. There must be an advantage to the programmer, not simply to the responsible manager who needs to somehow wrestle design documentation from those whose true interests lie elsewhere. Probably the best current methods to achieve this goal employ the proven concepts of structured analysis and design.

If the product being developed is of any size or complexity, it may be wise to plan on using

1. Technical writers for documentation whose purpose is to instruct.
2. Technical writers, and possibly cooperative development staff members, for reference documentation.
3. Staff members for producing design documentation, providing that a methodology can be agreed upon that is advantageous for programmers to use. Structured analysis and design techniques provide significant advantages in the development of good documentation in parallel with software development.

Facilities

In development, facilities include construction, acquisition, or conversion of buildings for laboratory space, prototyping, training, testing, utilities, office spaces, environmental control, and so on. This is also the point in the logistic support WBS to include procurement of any furniture, copiers, fax machines, telephones, lamps, staff computers, vehicles, perishables, and so on needed to support development, in addition to available existing equipment and facilities.

Operational facilities include the development, acquisition, or conversion of structures, vehicles, and so on that have not been specifically designated in the "auxiliary equipment" category of the mission product. A guideline for differentiation between these two categories is that auxiliary equipment is generally oriented toward the support of a mission product item, or subsystem, while operational facilities refer to systemwide support facilities. A mobile field system may include a number of internal systems, each of which is mounted on its own vehicle. Thus a vehicle on which a communications center subsystem is mounted may be classified as auxiliary equipment to that subsystem. Another vehicle that provides service to all subsystem vehicles may be classified as logistics support facility. There are always options to some extent, but consistency is to be valued. The important thing is to think of everything and to recognize that, in a good generic WBS, everything has a well-defined place.

I recall an incident while working as a communications consultant for a public safety agency and assisting, among other things, in facility design.

"Don't forget the chairs," I was warned.

"The chairs?"

"Yeah, the last time we did this we were so hung up in communications equipment, system throughput, queueing theory, and consoles that we forgot the chairs. Did you know that the chairs we use in the command and control center cost close to $500.00 apiece and we use thirty of them? That's a lot of money for us to forget. Don't forget the chairs."

Errors of omission in consideration of facilities for support of development and operations are among the most common. Before you are done with planning for facilities, take a mental walk around and picture yourself working there.

Personnel and Training

On the development side, this item basically covers the time and money required to acquire and train new personnel. It can easily take two to three months, and may require up to eight months, before newly acquired personnel are totally up to speed in their productivity. If new personnel are required, this logistics support item will affect both costs and schedules.

The design of training for operations personnel can have at least two aspects. In some systems, initial training in operations and maintenance at the time of system delivery is all that is contracted for. In others, such as in the maintenance of consumer products, design of a sustaining training capability is required as part of the system logistics support design. While related, the requirements for these two approaches differ in depth. The goals of the training component of logistics support need to be well articulated, as does everything else, in the early establishment of system requirements. Training for operations includes training services, equipment, hardware and software aids, test equipment, and parts sufficient to impart the necessary skills for operation and maintenance of the mission product. Included is the development of derived requirements and design and implementation for all training equipment, as well as the execution of the training function.

A final word on logistics support. Before final acceptance or delivery of any system or product, all facets of logistics support *will take place—even if planning for logistics support is totally disregarded.* Various events happen because they have to happen. The only question is, how well were they anticipated? Remember me at the Texas gate? Every element of logistics support will come up sooner or later because there are no products that do not need support of one type or another. If planning has not been adequate, even down to the logistics of arriving at a gate with a delivery, then a reactive mode must be assumed involving adjustments, afterthoughts, and unscheduled time and money.

THE RESOURCE PLAN

Resources consist of money, time, workforces, contractors, facilities, equipment, and materials. Resource planning for product development is the process of assigning available resources in a logical and ordered sequence for the purpose of executing a product development. Results of successful resource planning are schedules and budgets that can support successful product development, and that meet, or fall within, resource constraints.

Development of a resource plan entails the following steps:

1. Development of work precedence charts. Work precedence charts show the sequence in which the creation of products must take place, as well as the dependency relationships between products.

2. The assignment of time lines for the execution of interim product development for each product in the work precedence chart in order to produce schedules.

3. The assignment of resources to each time line to accomplish system development resulting in a budget.

4. Iteration of steps 2 and 3 until a plan emerges within the given resource envelope developed during preproject planning.

Work Precedence Charts

Work precedence charts are constructed using one of two approaches. One approach graphically displays activities associated with arrows that connect nodes. The activity-on-arrow format is often used for the critical path method (CPM), the critical path scheduling (CPS), and the program evaluation and review technique (PERT) methods for showing activity precedence. The second approach graphically displays products within nodes and shows the precedences for product development.

This second approach to work precedence charts should be used for resource planning because it is product oriented. This allows the maintenance of correspondence with the work breakdown structure. The process of resource planning typically provides further insights into interim product definitions—hence, an opportunity to update details of the work breakdown structure. *The two should always faithfully correspond.*

Logically, there can be four combinatorial precedence relationships between the creation of two products, A and B. They are

- Finish A to start B—Work on B cannot start until A is finished.
- Start A to start B—Work on B cannot start until work on A starts.
- Finish A to finish B—Work on B cannot be finished until A is finished.
- Start A to finish B—Work on B cannot finish until work on A starts.

The finish-to-start relationship can handle the great majority of situations. The last relationship, start-to-finish, while often mentioned for completeness, is somewhat rare and of less use.

There are a myriad of graphical conventions used to depict these precedence relationships. A far simpler scheme is to recognize that the basic issue deals with whether products must be developed serially or whether their production may overlap in time. For example, a detailed design typically cannot begin until functional requirements are concluded. Production of documentation for detailed design, however, may overlap the design effort. Clearly, such parallel representations of interim product developments involve a great deal of interaction between the two efforts. In this scheme, products either have a serial dependency or they may be realized in parallel. The parallel configuration can be extended to indicate that activity B may start after activity A has begun by graphically overlapping activity B to the right of activity A.

The process of work precedence charting should begin with the top-down representation of work breakdown structure items in precedence chart form. The representation should be carried completely to the bottom. The bottom is defined as the lowest-level work breakdown structure items for which it is desirable to define products and/or financial accounts—that is, the lowest level for which programmatic and technical resource management practices are deemed mandatory. Along the way, there must be a freedom to add useful products as additional practical insight is gained. In particular, the bottom levels must be considered anew with great care in the interest of completeness and consistency. Completeness addresses the issue of whether all needed interim products have been considered for the realization of *each* product. Consistency addresses identification of receivables needed from, and deliverables owed to, other work breakdown structure entities and the timeliness of their precedences. When, and only when, it is felt there is completeness and consistency at the bottom are the charts then iterated upward to include additions and corrections. The process continues top down and bottom up until a complete and consistent set of work precedence charts is developed.

Each of these products may, in turn, be broken down further into other subordinate products. When the bottom is reached, all needed receivables to accomplish the realization of each product are added. Interaction with other work precedence charts under development at the same level helps define deliverables needed by other corresponding efforts. These are also added. Both receivables and deliverables must be agreed upon by all parties involved at all affected levels.

All charts, now including receivables and deliverables, are iterated upward to the end product level. The process is repeated downward and upward until a realistic, complete, and consistent set of charts results.

Time Line Assignment

Following the construction of work precedence charts, the assignment of time lines for the development of intermediate products takes place from the bottom up. The addition of time lines to the work precedence chart creates schedules. Time lines are best developed by basing estimates on historic data for similar product developments and on comparable experience. Estimates should not be influenced by the amount of time that is believed to be available, but rather on honest and realistic judgments.

If time lines emerge with resolutions of less than a week, the product definitions should be revisited. The managing of resources with too many specifics can become a significant burden. On the other hand, time lines of two to three months should also be avoided in order to maintain sufficient visibility into actual progress. The time line detail should roughly approximate reporting periods.

A common method for clarifying time line interdependencies uses the program evaluation and review technique referred to as PERT. PERT was originally developed in 1958 for the Polaris missile project under U. S. Navy sponsorship. The approach is summarized in Figures 5.27 through 5.30.

Results of the first step in the use of PERT are shown in Figure 5.27, where the predecessor activities for each required activity are listed. Also listed are estimates for the most optimistic times (T_O), or shortest times, to complete each task along with the estimated mean times (T_M) and the most pessimistic, or longest, times (T_P), to completion for each task. The first activity in the sample schedule, activity B, is the construction of a facility. The execution of this activity does not depend on any other activity. Activity F, that of safety inspection, cannot begin until activity B is concluded. Similarly, dependencies for the other tasks are noted along with estimated values for T_O, T_M, and T_P.

This information is represented graphically in Figure 5.28. The values entered in the circles are arbitrary node reference numbers to facilitate discussion only. The execution of activities is represented by arrows between these circular nodes. Values for T_O, T_M, and T_P are indicated, along each activity path in parentheses. Each activity is directed to a unique node. Since activity A cannot begin until activities C, E, and F are all completed, it is convenient to note the completion of these three activities at the single node numbered 50. Dummy activities are used in PERT charts to effect this closure.

The next step is to develop an estimate T_E for the expected time to complete each activity. Figure 5.29 displays the results of the calculations for T_E for each activity path. The PERT convention for determining this estimate is given by

$$T_E = (T_O + 4T_M + T_P) / 6$$

which can be viewed as a mean value with weighting on the T_M term.

It is clear that the longest time path from node 10 to node 60 consists of passage through nodes 20 and 50. This path constitutes

ACTIVITY	PREDECESSOR ACTIVITY	TIME ESTIMATES (WEEKS)		
		To	Tm	Tp
B. BUILD FACILITY	NONE	20	24	30
F. SAFETY INSPECTION	B	2	3	4
C. INSTALL EQUIPMENT	B	8	16	20
D. RECRUIT WORKERS	NONE	2	2	3
E. TRAIN WORKERS	NONE	4	5	6
A. PERFORM PILOT	C, E, F	4	5	9

Figure 5.27. Pert Example.

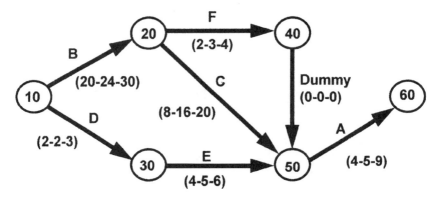

Figure 5.28.   Pert Example (Cont.).

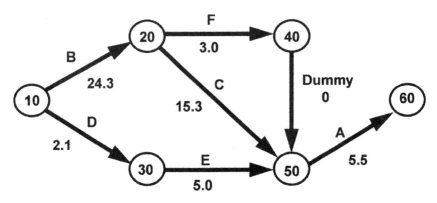

Figure 5.29.   Pert Example (Cont.).

the critical path for the entire project. Figure 5.30 highlights this critical path and also notes derived values for the earliest expected times,
TE, and the latest expected times, TL, that an activity can be expected
to take to reach the node toward which it is directed. For example,
the earliest expected time that activity completion to node 40 can
occur is just the sum of the expected times for activities B and F, or
27.3 = 24.3 + 3.0. Note, however, that activity F does not really need
to be completed until the critical path activities of B and C are completed. This is because activity A, the last activity, cannot begin until
activities F, C, and E are all completed. This means that activity F can
take as long as activity C without jeopardizing the schedule—an
amount of time equal to 15.3 time units. Thus the longest time, TL,
that can expire in reaching node 40 is 39.6 = 24.3 + 15.3. Figure 5.30
also shows similar calculations for the remainder of the network. The

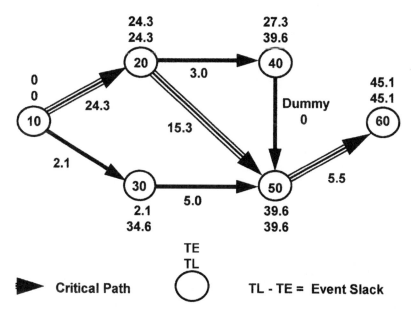

Figure 5.30. Pert Example (Cont.).

difference between a given value for TL and TE (TL – TE) is referred to as the event slack. Note that the event slack along the critical path is simply equal to the expected time along the critical path; that is, there is no extra time to spend.

The PERT tool is useful in bringing a deeper understanding to the time line development process. It is also clearly useful in pointing out those activities that require the most management attention and in providing a valuable tool in identifying high risk areas.

After time lines are constructed at the lowest levels, they are then merged upward through each successively higher level until the end product level is reached. Conformance with the overall product development time line requirement is then examined. Time lines are then adjusted downwards to conform to the top-level schedule envelope. Iterations continue until all schedules are complete and consistent.

With a time line completed, a receivables and deliverables database should be constructed and maintained to include

- product name
- delivering work breakdown structure entity
- receiving work breakdown structure entity

- available date of product
- required date of product
- a comments column

The status of receivables and deliverables should be routinely covered, along with the status of all other products, at each reporting period.

Resource Assignments

Following the assignment of time lines, or a schedule, within the top-level system development schedule envelope, the assignment of resources for the development of products takes place from the bottom up.

The first resources to be assigned are workforces, contractors, equipment, and materials required to accomplish the development of each interim product at the lowest levels within the previously developed time lines, that is, schedules.

As with time lines, resource loads are best determined through experience with similar developments. If experience within, or outside, the project is not available, then best guesses must suffice.

Cost estimates are then assigned to the physical resources to create a budget. Costs can be grouped into four categories:

- work force
- support services (such as configuration management documentation, printing, copying, and so on) that are charged to the development effort
- procurements including parts, subcontracted products, consultants, computers, software, and the like, used to directly support development
- overhead to include costs for sick time, secretarial support, vacations, other benefits, routine supplies, and the like, not directly related to the project

Estimated costs are merged upward until a total system (end product) development cost is aggregated. If the total estimated cost does not conform to the cost envelope developed during preproject planning, then iteration is required. If cost iterations alone do not achieve the desired result, then the schedule may need to be iterated as well. The changing of time lines will, of course, impact estimated costs. The adjustments are clearly interdependent, and many iterations may be required. Adjustments may eventually be required in the original preproject product concepts, as well as in parts of the project plan developed to date. Eventually, a middle ground is found, or not found, and a decision is made either to embark upon implementation, or dump the whole thing.

The Resource Planning Team

On developments of any reasonable size, resource planning should be done only by experienced personnel. Segment, element, and subsystem managers and cognizant engineers (i.e., technical personnel) are typically not sufficiently trained in resource planning, nor are they usually organized to accomplish this effort in a coordinated manner. It is important that the resource planning effort be organized and managed by knowledgeable members of the project staff. If extensive experience is not present within the project staff, then this capability should be acquired and costed as an additional service, or a procurement. In either case, it is clearly a part of the project staff work breakdown structure.

Technical personnel are, however, the best resources for estimating time lines and costs. Therefore, resource planning should be organized and managed by the project staff but should call upon technical personnel, when needed, for data estimates. Technical personnel should be briefed on the generic approach and made aware of their specific contributions. With this approach, a coordinated, cohesive, and consistent resource planning effort can be carried out across and between all levels. Technical personnel should not need to interact with the resource planning project staff more than two to four hours a week.

THE PRODUCT DEVELOPMENT TEAM MANAGEMENT PLAN

The Concurrent Product Development Team Management Plan is the top-level technical management plan that states how the goals of the concurrent product development team are to be met for a specific project. Construction of a comprehensive plan represents a valuable opportunity for the development team leader to define the technical structures and processes to be employed in the complete execution of concurrent product development and team member responsibilities. The plan must be consistent with, and supportive of, higher-level management plans. The document should consist of a series of succinct paragraphs, not a magnum opus, which state the strategies to be employed in key areas of technical management tailored to the development at hand.

▯⁝ **TABLE 5.25**   *Suggested Topics for the Product Development Team Management Plan*

1. Organization
2. Concurrent Product Development Team Paradigm Selection
3. The Product Development Team
4. Software Analysis and Design Techniques
5. Product Development Priorities
6. Margin Management
7. Options Analysis
8. Logistics Support
9. Configuration Management
10. Performance Measurement
11. Risk Management
12. Test and Evaluation
13. Schedules

Suggested topics for inclusion in the plan are listed in Table 5.25. The following briefly describes the scope and content for the treatment of each section in the plan.

Organization

This section should include brief text supporting an organization chart showing the reporting relationships between the project manager, project staff, the development team leader, the software development team leader, cognizant engineers, concurrent engineering teams, and the system user organization. The relative positions of the development team leader and software development team leader should also be included. The team organization is of critical importance. Many product development failures can be directly attributed to structures that are incapable of properly executing concurrent product development team functions.

Paradigm Selection

This contains a brief statement of the rationale for the selection of either the staircase with feedback, early prototype, spiral, or rapid development method models, or a combination thereof.

The Concurrent Product Development Team

This section specifies the permanent membership of the team by title and specifies temporary membership, by discipline, that will be called upon to support the team, including required disciplines in specialty and concurrent engineering. This section also states the meeting frequency of the team, specifies any meeting location rotations (implementor, user, contractors), if appropriate, and establishes the roles and responsibilities of the team.

Software Analysis and Design Techniques

Under this topic, the strategy for use of analysis and design methodologies to support software development is discussed. Software development methodologies include structured analysis and design techniques incorporating data flow diagrams and structure charts, object-oriented programming, program development folders, and so on.

This section should include any plans for analysis of the utility of computer-aided resources and design techniques in support of development. The plan states that the use of these tools will be investigated, and selections will be made by the team under direction of the software development team leader and approved by the concurrent design team.

Product Development Priorities

This section lists the product development priorities, defines the rationale for each, and describes how they will be used as top-level criteria for design trade-offs, requirements flowdown, and related technical and programmatic decisions.

Margin Management

The candidate commodities for which margins will be established and maintained are identified. Also presented in this section are margin management philosophies, methodologies, and values to be established at PRR, PDR, CDR and other appropriate points in the development as may be desired. Remember, the philosophy for margin management is closely structured by the selected concurrent product development team paradigm.

Options Analysis

Here the plan identifies the points in the development cycle where any need for resolution of requirements and design issues is anticipated. This section of the plan also describes appropriate methodologies such as early prototyping, analytic modeling, dynamic modeling, static loading, mockups, physical analogues, bread boarding, use of a series of operational deliveries, and so on, that will be used to gain confidence in proposed options and control risk. Again, the structure of the options analysis strategy is influenced by the concurrent product development team paradigm in use.

Logistic Support

This section of the plan describes the top-level strategy for implementation of logistics support to include discussion of each major logistics support item that needs to be addressed in the design of the product. Depending on the project, major items may or may not include concepts for maintenance planning, supply support, test equipment, technical data packages, transportation and handling, facilities, and personnel and training. A succinct description of the operational maintenance philosophy is included here, describing how the system will be maintained and by whom. A similar description of major logistics considerations to support the development of the product, as required, is also included.

Configuration Management

The basic approach to configuration management as it will be applied to the project is stated. This section should also clarify the extent and timing of control as a function of the concurrent product development team paradigm to be used. Baseline freeze points are defined for requirements, system specifications, internal system designs, reviews, and other baselines as required. This section also defines how a set of baselines will be established in accordance with the chosen paradigm.

Performance Measurement

The tools and processes by which the development team leader plans to maintain visibility over the development process are defined in this section. The discussion specifies which tools—such as Gantt charts,

PERT charts, low-level schedule monitoring, reviews, readings, walk-throughs, early testing, earned value determination, and so on—will be employed for the technical monitoring of system and internal system development.

Risk Management

Any anticipated potential high-risk issues likely to be encountered throughout the development cycle are identified here. The risk management approach and timing to be employed, which is basically driven by the concurrent product development team paradigm in use, is also defined. The top-level technical back-up design and implementation strategies to be used in the event of failure of initial development efforts and the timing for use of backup strategies are defined as well.

Test and Evaluation

This section, at a minimum, discusses the top-level sequence strategy and methods of testing for parts testing, for unit testing, for integration testing, and for product-level testing. Other testing strategies may involve the need for pre-ship and post-ship testing, acceptance testing, and early supplier involvement in testing. This section also states when and by whom the test plans and procedures and detailed test plans will be developed, as well as who will do the testing, as a function of the concurrent product development team paradigm employed.

Schedules

This section presents a top-level schedule on a single page for the major concurrent product development team deliverables in a manner consistent with the adapted concurrent product development team paradigm for the project. In the generic concurrent product development team process, the plan is generated by the development team leader immediately after the formation of the concurrent product development team—that is, early in the overall processes. The plan is then reviewed by the concurrent product development team, iterated as required, and submitted to project management for approval.

DETAILED ROI

The detailed return on investment (ROI) analysis provides a more thorough estimate of the time envelope in which cash flow becomes positive. This time line includes development time. As the venture capital guys say, "When is the nut?" It also includes a refined guess as to the rate of growth of cash flow based on market share estimates.

Phase B ROI is best built by finance and marketing planning team members. The ROI at this point can be more detailed because there is a much better understanding of development costs than there was at the end of Phase A. The tricky part remains estimating future market share and associated revenue. Unless you are privy to new information, there isn't much you can do other that sharpen up the cash flow estimates made during Phase A. Call on the experts in your organization. Beware of being too optimistic. Be careful and know what you don't know: This is very important; it's another good reason for teams.

PRODUCT DEVELOPMENT GO/NO-GO REVIEW

The last step in the project planing process is to obtain approval for implementation. Every element of the finalized implementation plan should be reviewed for completeness, realism, and consistency with the previously approved preproject plan. This is a crucial review board meeting and may easily last for more than one day. Approval at this point represents a total organizational commitment to full-scale system development.

The presentation may run from half a day to a few days, depending on the complexity and how familiar management is with the product. Backup material should also be available in the form of additional slides for each item covering the entire output of the team's efforts. That is, as in the Phase A presentation, the main presentation should not be overly lengthy or cover detail that the management team may not need or want. The main presentation should be crisp, complete, and to the point. The backup slides are pulled out when elaboration is specifically requested.

Here is a guideline for the content of the main presentation:

Topic	Number of Slides
Purpose	1
Content	1
Finalizing user needs	3
Building a work breakdown structure (WBS)	2–10
Organization	1–2
Control policies	
Mission statement	1
Product development priorities	1

Review and reporting structures and formats	5–8
Technical margin management	1–2
Documentation	3–6
Risk control	1–2
Configuration management	1–3
The Logistics Plan	2–5
The Resource Plan	1–3
The Product Development Team Management Plan	1–3
Detailed Return on Investment (ROI) Plan	1–3
Recommendations	1

Assuming you have passed this review and received a "go ahead" for full scale product development, you are ready for Phase C.

ENDNOTES

1. B. Boehm, "Software Requirements Analysis and Design," Data Processing Management Association Seminar, San Jose, CA, July 1985.

2. "Systems Engineering Management Guide," Lockheed Missiles and Space Company, 1983.

3. Ralph Keenly, and Howard Raifa, *Decisions with Multiple Objectives* (New York: John Wiley, 1976), and Ralph Keenly, *Value-Focused Thinking* (Cambridge, MA: Harvard University Press, 1992).

Part

6

Phase C: Concurrent Product Development

■■■■■■■■■■■■■■■■■
••

GENERIC PHASE C ISSUES

This section presents a generic overview of Phase C issues. As with the material discussed in Phases A and B, the intent is to aid in the attainment of completeness by covering the generic set of issues in the interest of avoiding undue surprises. In support of the implementation effort, modern systems engineering techniques for definition of requirements, requirements flowdown, conducting of trade-off analysis, interface definition and control, and construction of test plans and procedures are also presented. These subjects, which directly support product development, will be covered following an overview of the complete Phase C process and a discussion of who is on the product development team (PDT).

Phase C Overview

Phase C basically implements the plan constructed in Phase B. Thus, we will call upon our updated understanding of user needs, the adopted work breakdown structure, and our plans for organization, control policies, and logistics, as well as our resource plan and the Product Development Team Management Plan.

The generic staircase paradigm product development process is summarized in Figures 6.1 through 6.4. The first three figures show the steps carried out in preparation for each of the reviews planned for in Phase B. The fourth figure, Figure 6.4, shows the generic steps carried out for physical construction of the product. The process depicted in the figures assumes that Phase B planning called for at least three formal reviews; a product requirements review (PRR), a preliminary

223

224

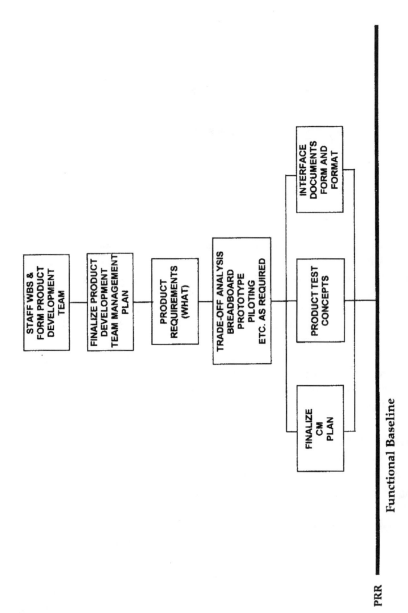

Figure 6.1. The product development process for the staircase paradigm.

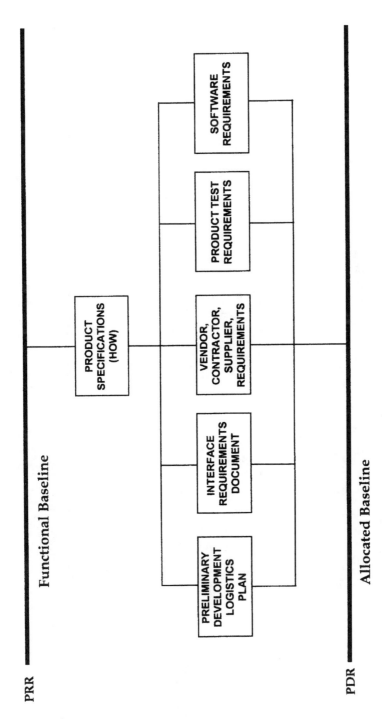

Figure 6.2. The product development process for the staircase paradigm (Con't).

225

226

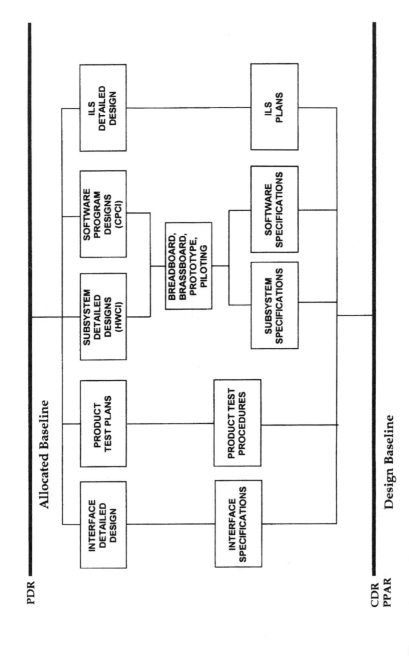

Figure 6.3. The product development process for the staircase paradigm (Con't).

CDR
PPAR   Design Baseline

Non-Production
route

Production route

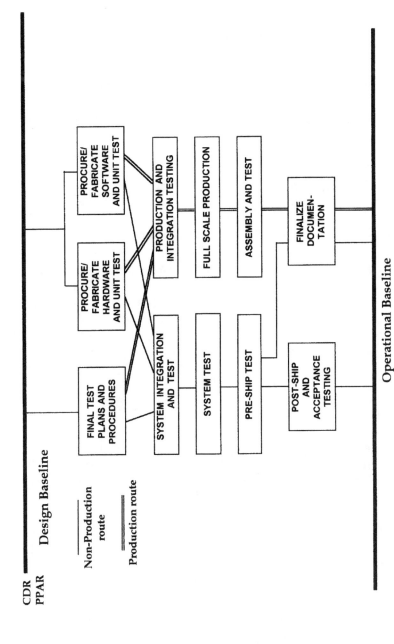

Operational Baseline

Figure 6.4.   The product development process for the staircase paradigm (Con't).

227

design review (PDR), and a critical design review (CDR). This is a minimum set of reviews for any reasonably complex product. If you elect to have more reviews, or gates, then decide the exact purpose of each within the context of the overall process and insert them at the appropriate points tailored to your needs.

The major differences between the staircase development, the early prototype, and the spiral models is a matter of timing in the finalization of requirements. The staircase model should be used only when requirements can clearly be articulated by the user community, or by valid representation of the user community, and required technologies are well understood.

Early prototyping essentially extends the trade-off analysis step shown in Figure 6.1 to include more detailed models for the purpose of understanding requirements. These models may include breadboarding, brassboarding, entire mock-ups, creation of workstations, construction of mules, pilots, and so forth. When requirements are understood, we can then revert to the remainder of the staircase paradigm.

The spiral model was first developed for software-intensive products and systems. In this scheme, requirements are not completely finalized until the trade-off studies shown in Figure 6.3 are concluded—that is, during the detailed design. This paradigm requires excellent planning and understanding of all who are involved. Exceptional management skills are required, particularly in risk management. The spiral model is useful when trade-offs between requirements and cost are unclear and cannot be resolved until the design is carried out in significant detail. This means requirements must be capable of change late in the process because there is a greater risk with regard to implementation strategies. It is not a comfortable paradigm for most people to use because some of the "Whats" can be driven by the "Hows." Constant customer involvement is imperative when using this paradigm, and the requirements envelope must have some flexibility. Configuration management and review content must be tailored to each use of the model.

The rapid development model is different from any of the others. Here the staircase concept is modified slightly and repeated. When using this model, follow the guidelines covered in the Phase A discussion of paradigm development.

The major generic steps that should be carried out in Phase C consist of

1. Staffing of the product development team and the WBS
2. Requirements development

3. Preliminary design
4. Detailed design
5. Implementation following the critical design review
6. Testing

The first step is to select the product development team (PDT) leader if this was not already done during the planning Phase B. Inclusion of the PDT leader during Phase B planning has the significant advantage of receiving his or her inputs during this most important stage and promoting the all-important component of ownership. The team leader should also be heavily involved in the staffing of the PDT and the WBS.

The generic steps in requirements development are depicted in Figure 6.1. The first meetings of the PDT are devoted to review and finalization of the Product Development Team Management Plan. This establishes understanding and concurrence among all team members on how the team plans to do business.

The team then begins development of formal product requirements. This may, or may not, involve conducting trade-off studies to deal with any questions regarding feasibility of requirements themselves as they emerge, at any level. Sections of the requirements document are assigned to PDT members in accordance with the requirements document outline. For example, systems engineering deals with the systems view, cognizant subsystem engineers work on the subsystem sections, logistics people on logistics, and so on. Remember, if you are generating a pure requirements document first, the content consists of measurable whats, not hows. The hows come later in specifications. In any case, the whole point of generating requirements at this point is to show that the team has a thorough grasp of what the customer needs and wants in preparation for the product requirements review (PRR). At the PRR, the team should also be prepared to present how it intends to exercise configuration management, how it intends to verify that the product works (testing concepts), and the form and format for interface documentation. Details on testing plans and procedures and on interface development will be presented later in this section.

The basic purpose of the product requirements review is to demonstrate that the user, or sponsor, requirements are understood *and* that a feasible approach to meeting these requirements is understood. Passing the PRR gives permission to the PDT to proceed with a preliminary design for the product. At this point in time, the development process has reached what is commonly referred to as the functional baseline.

Having passed the PRR, the PDT now focuses on the hows by developing product and subsystem specifications. We are on the road to the preliminary design review (PDR). In addition to developing specifications, PDT members under the leadership of the team leader are now developing the following:

1. A first cut at a logistics plan to support the development process. (Remember that logistics requirements for operations, when the product is released, are a part of the product requirements and specifications.)

2. Production of an interface requirements document. (More on this later.)

3. A statement of vendor/contractor agreements and interactions. These include adherence to quality standards.

4. Development of product test requirements that have been derived from the functional requirements document. (Each product level requirement needs to be tested. That's why top-level requirements must be measurable.)

5. Documentation of software requirements. (Note that the product specifications state how the product is to be built. That is, it states what features are to be realized through hardware and what is to be implemented through software. While much of this may be obvious early in the game, the distinction is not really pinned down until the preliminary design step of formalizing product specifications.)

Each of these topics is covered at the preliminary design review (PDR). The purpose of this review is to demonstrate the feasibility of the chosen design. Requirements have been allocated to specific hows. Hence, upon completion of the PDR, the design status is referred to as the allocated baseline.

Having passed the PDR, the PDT turns its attention to the items shown in Figure 6.3. This is the detailed design phase that results in interface specifications, complete product test plans and procedures, subsystem and software specifications, and the logistic support plan for development. Further design trade-off studies may be carried out to resolve any remaining questions regarding the achievability of the detailed design. Note that nothing has actually been built yet. We are ready to seek approval to do so at the critical design review (CDR).

The purpose of the critical design review is to demonstrate that the design is mature and ready to implement. A production part approval review (PPAR) may also be a part of the CDR to assure that vendors are ready to be cut loose. Production part approval, when used, is a quality-oriented program that continues throughout development. If we pass the CDR, we can actually cut metal, begin coding, and turn on any procurements for hardware and software. We have reached the design baseline.

The basic steps from CDR to product delivery are shown in Figure 6.4. The process is slightly different for one-of-a-kind systems as opposed to products slated for production. The one-of-a-kind path is shown by single lines and the latter by double lines.

For one-of-a-kind products, after a successful CDR, hardware units and software are procured or fabricated and unit testing takes place. This is followed by system integration and testing, system-level testing, pre-ship testing, post-ship testing, and customer acceptance testing.

For products bound for production, after a successful CDR, hardware units and software are procured or fabricated. This is followed by any required production and integration testing on early serial numbers to reduce risk. Full-scale production can now take place followed by assembly and test operations.

WHO'S ON THE PHASE C TEAM?

The following disciplines, as a minimum, should be represented on any development team on a permanent basis:

1. Systems engineering
2. Cognizant engineering for each product subsystem
3. Software PDT leader
4. Integrated logistics support engineering
5. Test engineering
6. User representation
7. Production
8. Assembly
9. Other concurrent engineering team members, as required

Of course the Product Development Team leader is the chairperson.

Systems engineering is represented by the PDT leader. Larger projects may involve a complete systems engineering staff. In these cases, the systems engineering staff leader should also be on the team.

On larger projects, a single cognizant engineer (COGE) is warranted for each of the mission product items, that is, subsystems. In our automobile example this would include a manager at the power train, chassis, and so forth levels. If the new product is toothpaste, there would be at least three subsystems, or disciplines, required. One is the package. The package is important because it must preserve the contents and have a way of getting the contents out. Other packaging issues may include safety, size, and so forth. A second discipline (subsystem) involves the ingredients of the package content. Just think of the possible issues involved regarding the content. What is the mission of the product? Is it designed for a niche market such as children, seniors, or veterinarians? Is the product expected to be viable for a relatively short time, or for decades? What is it supposed to do for your teeth? Should it have a flavor? The chemistry of toothpaste is a discipline in itself. A third discipline involves the text that goes on the package. What colors are to be used, and the like? Clearly, marketing plays a role here, too. The point is that a number of disciplines are involved, and someone has to be accountable for them, to manage them. (Notice something here. I don't really know much about toothpaste, but look at the reasonable start that can be made just because we have a structured way of thinking about it.)

On software-intensive projects, cognizant software engineers may also be included. Alternatively, the system software engineer may run his or her own software-concurrent product development team and report on software activities to the PDT.

On smaller projects, a single engineer may have cognizance over more than one mission product item or work breakdown structure item. For example, integrated logistics support (ILS) may be absorbed into the PDT leader's role. I worked on an aircraft once where I was cognizant engineer for two subsystems—an inertial navigation subsystem and an infrared sensor subsystem. This is OK. All items, however, should be represented.

Integrated logistics support and test engineering are other necessary major functions for representation. This representation ensures that important design impacts driven by these functions are not overlooked.

The presence of the user, or faithful user representation, on the PDT cannot be overstressed. *A major cause of product failures lies in the construction of systems that simply do not do what the customer wanted. A concurrent product development team cannot be successful without constant customer representation.*

There is no question that operation of a PDT without the continual presence of the customer, or faithful customer representation, is *extremely* dangerous. The holding of baseline reviews for the customer, while necessary, is typically not enough. Constant representation of the customer's view is crucial, particularly in the development of products for new or unique applications. Customer representation is also crucial in products designed simply to provide major improvements over existing systems. In these settings, requirements can change as insights are gained through the development process. Engineers are notorious for solving problems with the most expedient "engineering" solutions, which may, or may not, be customer oriented. When left alone to cope with the myriad of issues that arise throughout the development cycle, the PDT leader can absolutely count on going astray of user needs in the absence of adequate user feedback.

This is not to suggest that the customers can constantly change their minds with regard to fundamental direction. Baseline commitments are seriously maintained. Rather, the customer is there to maintain the development team's touch with reality. The customer is there to guard against building a product that cannot be used or that is not even wanted. The fact that exactly this happens in an alarming number of instances is warning in itself.

Most vendors, unfortunately, don't like continual customer participation in their design activities and decisions. This mind-set, often driven by proprietary reasoning or by a simple "dirty linen" defense, is perfectly understandable. The history of product and system implementations clearly indicates the inherent dangers of development in isolation. When change is truly required, it must be identified and accommodated in a timely manner. The probability of success when the customer is represented on the concurrent product development team is markedly increased.

Additional membership on the PDT may consist of one or more of the following functions as required:

10. Producability and manufacturability engineering
11. Reliability/availability/maintainability (RAM)
12. Human factors engineering
13. Safety engineering
14. Quality assurance
15. Environmental engineering
16. Finance
17. Marketing
18. Customer service
19. Suppliers
20. Other specialties as required

These functions should be oriented toward concurrent engineering and may be utilized on a full or part-time basis.

Phase C Methodologies and Guidelines

Now we will discuss a number of methodologies and guidelines for Phase C product development. These methodologies and guidelines are applicable not only to the staircase paradigm but to all product development paradigms. (The reader may wish to refer to Figures 6.1 through 6.4 throughout this discussion.)

The methodologies and guidelines to be discussed include the following functions:

1. Staffing the PDT and the WBS
2. Reviewing and familiarizing all members of the team with the Product Development Team Management Plan and the basic sequence of activities to be executed during development
3. Development of requirements (hardware and software) and requirements flowdown to specifications
4. Design trade-off study methodologies
5. Methods for interface development and documentation
6. Development of test plans and procedures

Staffing the Product Development Team (PDT) and the WBS

First we organize for implementation. We organize by staffing the WBS functions. We staff for the functions of management, systems engineering, product implementation, product test, and logistics support. Staff size will, of course, vary with product complexity. One person may be assigned more than one function. The point, as always, is that all functions should be considered.

The PDT membership is selected according to the previously given guidelines. Remember, you don't have to reorganize the entire corporation to form a PDT. If your organization is not used to this kind of operation, first be sure your management thinks it's a good idea and supports you. Then you need only visit and talk with appropriate supervisors of the various departments for the personnel you want, go over your plans and goals (not the least of which is getting to market faster with a quality product), explain the professional commitment required, and explain why that expert is needed.

The PDT leader must be technically competent across the board and have exceptional skills in human relations. Scheduling time for training never hurts and is advised if this is the first time out. Also, it will help a lot to have everybody read this book and understand what you have adapted for your own use and the general tone of the PDT setting.

Reviewing and Finalizing Product Development Team Management Plan

Among the first items of business for the PDT is the review of each item in the Product Development Team Management Plan. The plan may be modified to some degree, but its basic integrity must prevail, be finalized, and be agreed to before proceeding. The PDT leader is always open to suggestions and improvements. However, remember that the PDT is not quite a complete democracy. *In the end, the PDT leader has 51 percent of the vote.*

Development of Product-Functional Requirements

In a pure world, functional requirements state "What" a product must do and specifications reflect the detailed design of "How" the functional requirements are to be met. In practice, the two are often mixed in a single, or a set of, document(s). Systems engineering purists insist on a rigorous separation between the what and the how, arguing that it is dangerous to put any design detail ahead of a complete understanding of what must be done. In particular, mixing the two can freeze the ability to consider options at the system performance level if some design detail is arrived at too early. All of this certainly makes sense, and I think in general it is a good idea to decide what before how. The point is particularly well taken when building a novel product or system with a low degree of inheritance from existing products. In some cases, I myself have insisted on it. Typically, they have been large, new, one-of-a-kind products or systems. But there have been other cases where I have not. For example, in products where there is significant inheritance of subsystems, elements, or parts, a great deal of the design is already done and documented. There is little point in going through reverse engineering to create a lengthy "what" document.

Still, in the interest of completeness, we will discuss the development of functional requirements and specifications as separate undertakings.

The Functional Requirements Document

Development of the functional requirements document (FRD) represents the earliest stages in system design. Completion of the FRD implies a number of significant accomplishments. The most salient of these are the following;

- The FRD establishes the top-level commitment for the entire project effort to which all subsequent design and technical documentation must conform and be traceable,
- User needs and product constraints have been thoroughly assessed and translated into a set of requirements that are technically realistic and at the same time will allow the user's mission to succeed,
- All product functions in terms of "What" the product must do have been stated in measurable terms,
- If the staircase paradigm is used, the PDT leader and the PDT are prepared to freeze the contents of the FRD at the upcoming product requirements review. In the early prototyping, spiral, and rapid development models, functional requirements are frozen in accordance with the process for requirements development for each paradigm,
- System partitioning has taken place, which consists of the allocation of all system functions to functional areas and/or to internal systems (subsystems).

The process of FRD construction clearly entails the establishment of attainable requirements. This realism may require looking forward beyond the "what" level into specific design issues to ensure the feasibility of implementation. The extent of the forward look required to support the development of the final FRD is dependent on the development paradigm employed.

The need for realism also presupposes that the finished FRD may not necessarily meet all of the user's needs as stated in the user needs document. The user needs document is not a commitment to implementation. It is basically a complete wish list. The user needs document does, however, provide an important source from which the FRD is to be derived. It is the FRD that constitutes the first serious and highly structured commitment in the design and implementation process.

This point is visualized in Figure 6.5, User Needs vs. Functional Requirements. The stated user need space is typically larger than the ultimate functional requirement space. The development of the FRD often involves the paring down of the full range of user "wants" to a realistic core of functions to which the implementor is able to commit. This does not suggest that some controlled research and development may not be required. It does mean, however, that it is feasible to meet

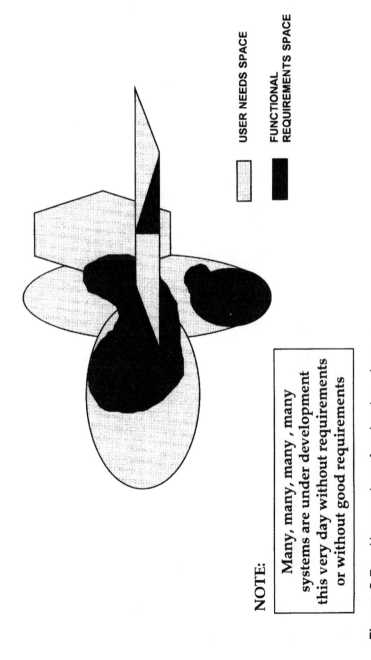

USER NEEDS SPACE

FUNCTIONAL
REQUIREMENTS SPACE

NOTE:

Many, many, many , many
systems are under development
this very day without requirements
or without good requirements

Figure 6.5.   User needs vs. functional requirements.

the functional requirements, consistent with the product priorities—
which may involve issues of performance, cost, and schedule. The PDT
forum, with user representation, is the principle arena in which this
negotiation takes place.

Every requirement in the FRD will eventually need to be vali-
dated. This means that each requirement must be measurable. Mea-
surability calls for the careful and complete definition of terms. Even
the most common and agreed-upon terms may mean different things
to different people. They typically do. If different interpretations are at
all possible, it is imperative to discover this now. For example, different interpretations of words can seriously set back progress during
testing and acceptance exercises.

It is useful to review examples of requirements that cannot be
measured versus some that can. The following are examples of
unmeasurable requirements:

- The personnel reporting program shall perform in real time.
- The system shall provide adequate memory for growth.
- The system shall provide a response time of four seconds for all
  nonadministrative messages.
- All terminal interfaces shall be user friendly.
- The system shall provide ample HELP files.

The following are examples of measurable requirements:

- The personnel reporting system shall be capable of completing
  a printout of individual personnel reports in one minute or less
  90 percent of the time.
- The system design shall provide for a memory margin of 100
  percent at the time of critical design review.
- The response time for the initiation of display of
  nonadministrative messages at user terminals shall be four
  seconds on the average and less than or equal to 20 seconds 95
  percent of the time.
- The top surface of all keyboard table heights shall be 23 inches
  above the floor.
- The system shall provide context-oriented help files for each
  executable function that may be initiated at the user's terminal.

In the first measurable example, use of the term "real time" is
completely avoided. It is a highly relative term whose scope is entirely
dependent on a particular problem environment. In one application,
real time can be microseconds and in another it could be hours. In a

fighting aircraft involved in target identification, acquisition, and engagement, real time can be milliseconds, even microseconds. In a medical setting, results of routine blood tests are often received the next day. Under such nonemergency conditions, the physician assumes that appropriate actions taken within 24 hours constitute a real-time response with respect to progress of the patient's condition. Also, the term "real time" in itself does not clarify the exact points in time in which real time is to be measured.

In the second example, the requirement for memory margin is restated with a measurable margin value to be realized at a specific point in time.

The third unmeasurable requirement does not sufficiently define response time in terms of the beginning of a message, the end of a message, or at what location in the system the transit of the message is to be measured. The restated requirement clarifies these points and also furnishes two statistical data points that are needed to characterize the service time density function in question. That is, specification of an average value alone does not fix the variance of the subject distribution.

Don't use the term "user friendly." The term, though popular, is not in itself meaningful, or measurable, in any consistent way, unless you want to hire a bunch of psychologists to come up with some answer. These issues may also need to be addressed through proper ergonomic requirements, which include those for maintaining posture, lighting, display distances and angles, response times, appropriate responses to user inputs, and so on.

Requirements for inclusion of *Help* files are common, but again, they must be stated in ways that allow unambiguous validation that the requirement has been met.

If you can't measure the requirement, don't write it. Requirements must always be realistic and measurable, but they must also be nonconflicting. For example, the requirement that an armored fighting vehicle be rapidly deployable to all terrain types along with the requirement that the vehicle withstand state-of-the-art antitank ordnance may be impossible to meet simultaneously. With current technologies, the first requirement demands a weight in the neighborhood of 20 tons and the second requires a weight in the 50-ton range. Deployability and survivability are clearly conflicting attributes. The user, however, may wish a design that addresses conflicting issues in the best manner possible. In these cases, it is useful to state acceptable ranges for conflicting requirements, instead of simple absolute values. This gives the designers and implementors a clear understanding of the extent of trade-offs that will be tolerated.

ATTRIBUTE UTILITY TRADE CURVES

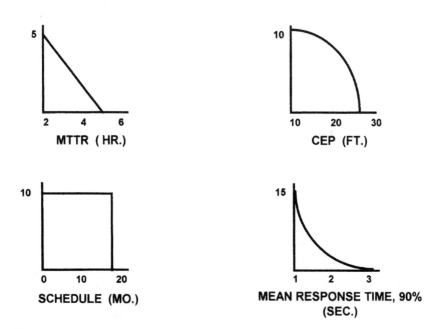

Figure 6.6. Dealing with conflicting requirements.

Utility Trade Curves   One useful way to convey requirement trade-off limits is through the use of attribute utility trade curves. Four examples of such trade curves are shown in Figure 6.6. The X axis values under the curves in the examples designate the acceptable numerical range for the requirement. The Y axis values give the relative importance, or utility, of meeting the requirement with a value in the numerical range.

The interpretation of the mean time to repair (MTTR) example in the figure is that 5 points will be awarded to utility for a MTTR of 2 hours. The utility linearly decreases to a maximum tolerable time value for a MTTR of 5 hours at which point the utility is zero.

The circular error probability (CEP) example establishes the tolerated range for the CEP from 10 to 26 feet. Attainment of a CEP of 10 feet is awarded 10 utility points. Attainment of a CEP of up to 20 feet is also acceptable. It is not until the design results in a CEP of over 20 feet that the utility rapidly decreases.

Attribute utility trade curves can also be used to convey the relative importance of programmatic issues. The schedule example in

Figure 6.6 shows that the system has no utility if it is delivered later than 18 months from the start. The curve also indicates that there is no increase (or decrease) in utility if the system is delivered at any point earlier than 18 months from the start.

In the system response time example, we see that 15 points are awarded if response time is less than 1 second for 90 percent of the time, and the points awarded diminish exponentially up to 2 and 3 seconds. When the attribute utility scheme is used, the FRD includes an overall system utility value that is to be met. The individual utility curves provide the implementors with a precise understanding of the extent to which design trade-offs can be made to meet the system utility metric. They also provide a means of quantifying the extent to which vendors propose to meet the requirement mix in the evaluation of bids and proposals.

FRD Generic Outline   A generic outline for a functional requirements document is shown in Table 6.1. The outline is meant to be a guide. Sections may be expanded, added, or deleted as appropriate.

The product definition section of the FRD provides an overview of the product's function and it's mission with supporting diagrams. Functional areas, or subsystems, are defined as part of this definition. (Characteristics and functional requirements of subsystems should be covered in Section 3.7.) Top-level product interfaces and interface requirements are described in terms of specific input and output products. Operational and organizational concepts refer to the relationship of the system to the overall mission, how it's operation accomplishes or supports the mission, characteristics of the direct users of the system, the system support mechanism concepts, and the structure of the overall user organization. Section 3.1 should not be verbose. It represents a top-level, self-contained, and comprehensive system description. It should be crisp, terse, to-the-point, and complete.

Product characteristics (3.2) are described with several subsections. Performance characteristics (3.2.1) include such items as acceleration requirements, fuel economy, message types, message formats, response times, data volumes, error rates, system life, storage requirements, turning radius, propagation coverage, ordnance delivery, and the like. These are measurable items directly related to performance of the final product. Physical characteristics cover weight limits, dimensional limitations, crew space requirements, ingress, egress, maintenance access, and so forth. Product level (reliability, availability, maintainability) RAM requirements are covered in Sections 3.2.3, 3.2.4, and 3.2.5. At the functional requirements level, it is common to state the system availability requirement only. This allows design trade-offs

▯⋮ **TABLE 6.1** *Functional Requirements Document Outline*

1.0 Scope

2.0 Applicable Documents

3.0 Requirements

    3.1 Product Definition
        3.1.1 Product Function
        3.1.2 Mission Descriptions
        3.1.3 System Diagrams
        3.1.4 System Interfaces
        3.1.5 Operational and Organizational Concepts

    3.2 Product Characteristics
        3.2.1 Performance Characteristics
        3.2.2 Physical Characteristics
        3.2.3 Reliability
        3.2.4 Maintainability
        3.2.5 Availability
        3.2.6 Environmental

    3.3 Design and Construction
        3.3.7 Security
        3.3.8 Safety
        3.3.9 Human Engineering

    3.4 Documentation

    3.5 Operational Logistics
        3.5.1 Facilities
        3.5.2 Transportation and Handling
        3.5.3 Personnel and Training
        3.5.4 Maintenance Plan

    3.6 Product Development Priorities

    3.7 Subsystem Characteristics

4.0 Constraints

5.0 Product Test Requirements

6.0 Appendices

between reliability and the maintenance support mechanism to be made at the specification (design) level of effort, thus providing maximum freedom to the designers. In cases where the maintenance and repair functions are not under the control of the developer, then it is appropriate to state reliability at the functional level. Bear in mind, however, that once two of the three system RAM parameters have been set, then all three have been set. Environmental requirements (3.2.6) include those for wind, rain, temperature, shock, motion, noise, electromagnetism, pressurization, blast, humidity, radiation, chemical, biological, nuclear, and the like. These describe the external conditions to which the other system characteristics will be subjected.

The design and construction sections of Section 3.3 include additional requirements for such items as security, safety, and human engineering.

A product is not a product without documentation. Section 3.4 states all requirements for documentation to be delivered with the system.

The logistics section (3.5) contains any known functional requirements not already covered related to facilities, system support, transportation and handling, personnel and training, and maintenance in the operational setting.

Section 3.6 contains a listing of the product, or system, development priorities, along with a description of their meaning and rationale for prioritization. This section can generally be taken directly from the Product Development Team Management Plan.

Subsystem characteristics are given in Section 3.7. There should be one subsection for each internal system (subsystem), covering performance and physical characteristics as a minimum. Sometimes separate subsystem requirements documents are created. If you do this, then follow the same outline, but do not simply repeat sections of the top-level requirements document.

Section 4.0 addresses implementation constraints. This is the only portion of the FRD that states "how" any part of the implementation is to be achieved. The need for this in the FRD may arise if, for example, a communication system must interface with existing computers such that a particular protocol must be used at given system interfaces. Products that are undergoing upgrades will commonly have constraints imposed upon them. Even brand-new products often need to make use of, or interface with, existing equipments. This section allows the authors of the FRD to instruct the implementors on these types of limitations.

System test requirements are stated in Section 5.0 of the FRD. Each functional requirement is stated in measurable terms so that it can be tested for conformance. This section also specifies how each requirement is to be tested. Verification methodologies consist of inspection, analysis, demonstration, and direct measurement. During system development, testing includes as a minimum unit testing, integration testing, and product-level (system) testing. Customer acceptance testing may also be included when appropriate. Pre-ship and post-ship testing may also be needed.

Many systems are too complex for complete testing by the user(s) prior to acceptance. If this is the case, it is important to understand this early on and to negotiate a set of acceptance test requirements with the user. Such tests typically are statistically based and should be included as a part of the developers' product test requirements.

Finally, appendices are included as required to support the FRD material.

Requirements Flowdown to Specifications

Requirements are developed from the top level and flow down in increasing and consistent levels of detail. But what is the top level? The top level consists of the mission statement and the project WBS. From these, we obtain product development priorities and the product WBS, that is, the breakdown of the product itself into subsystems. Figure 6.7 depicts the process and denotes top-level inputs and outputs of the flowdown process.

Let's apply this process to our Phoenix automobile example discussed earlier. Figure 6.8 lists the product priorities for the Phoenix. We begin by bringing initial characterization to each of these. Note that at this point some of the initial characterizations involve specific quantities and others do not. For example, a top-level requirement is to market in 2.5 years, and the retail cost is to be close to $12,000. With regard to appearance, comfort, and performance, the initial top-level requirements are less precise but still clear. We know we want a new body on this vehicle— something that hasn't been seen before. At this point, the comfort aspect is best characterized as better than, or as good as, the competition. And we want state-of-the-art vehicle performance. These concepts are derived directly from the mission statement.

Now we go one step further and look at how these top-level requirements distribute across the major product subsystems, the mission product top-level WBS. Figure 6.9 shows this initial distribution. Consider the schedule. In our example, the PDT decides to use an existing power plant. We will also use an existing drive/control system

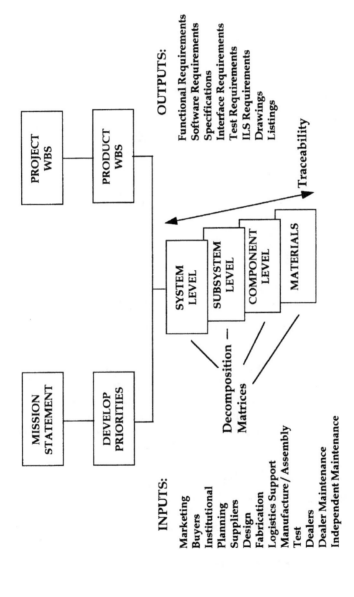

Figure 6.7. Requirements decomposition process.

INPUTS:

Marketing
Buyers
Institutional
Planning
Suppliers
Design
Fabrication
Logistics Support
Manufacture / Assembly
Test
Dealers
Dealer Maintenance
Independent Maintenance

OUTPUTS:

Functional Requirements
Software Requirements
Specifications
Interface Requirements
Test Requirements
ILS Requirements
Drawings
Listings

PROJECT WBS
PRODUCT WBS

MISSION STATEMENT
DEVELOP PRIORITIES

SYSTEM LEVEL
SUBSYSTEM LEVEL
COMPONENT LEVEL
MATERIALS

Decomposition — Matrices

Traceability

Automobile

PRIORITIES	PHOENIX
SCHEDULE	2.5 Years
COST	$12,000 base within 10%
APPEARANCE	New body
COMFORT	Better or as good as the competition
PERFORMANCE	Competitive sport coupe state-of-the-art

Figure 6.8.   Requirements decomposition process (Con't).

but we recognize there may be some modifications required. We are still at a high level here. The platform and interior will be new, but we still want to inherit as much as possible in these areas. It is clear at this point that this is where we will be spending the most money on development, and this is reflected in the cost row. The entries in Figure 6.9 represent top-level strategies across subsystems for the product development, and these will be further crystallized into firmer, measurable requirements as we go along. Significantly, each entry throughout this process is directly traceable to our priorities and to precisely what it is we are developing. This approach provides a logical structure for further definition. PDT members are asked to provide further details through assignment of specific action items.

Figure 6.10 reflects the next level of response. Now we are beginning to put numbers on schedules derived from the planning phase and from team updates. We are beginning to put numbers on cost and performance and finalizing strategies for developing the all-important

Automobile

PRIORITIES	POWER PLANT	DRIVE/ CONTROL	PLATFORM	INTERIOR	
SCHEDULE	Inherit	Inherit/ Modify	New/ Inherit	New/ Inherit	---
COST	Low	Low	Medium	Medium/ High	---
APPEARANCE	---	---	New	New	
COMFORT	---	---	Size, Suspension	New/ Inherit	
PERFORMANCE	SOA	SOA	New/ Inherit	New/ Inherit	

PHOENIX SUBSYSTEMS

Note: Top level strategy

Figure 6.9. Requirements decomposition process (Con't).

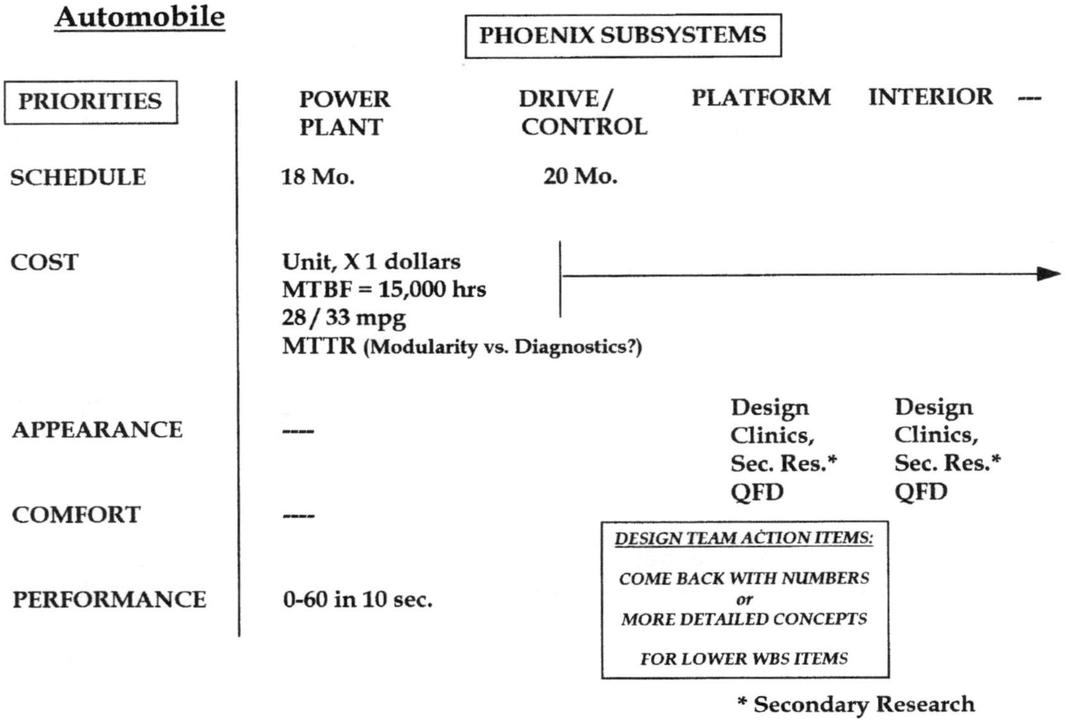

Automobile

PRIORITIES	PHOENIX SUBSYSTEMS				
	POWER PLANT	DRIVE/ CONTROL	PLATFORM	INTERIOR	
SCHEDULE	18 Mo.	20 Mo.			
COST	Unit, X 1 dollars MTBF = 15,000 hrs 28 / 33 mpg MTTR (Modularity vs. Diagnostics?)				
APPEARANCE	—		Design Clinics, Sec. Res.* QFD	Design Clinics, Sec. Res.* QFD	
COMFORT	—				
PERFORMANCE	0-60 in 10 sec.				

DESIGN TEAM ACTION ITEMS:
COME BACK WITH NUMBERS
or
MORE DETAILED CONCEPTS
FOR LOWER WBS ITEMS

* Secondary Research

Figure 6.10. Requirements decomposition process (Con't).

appearance requirements. Note that the power plant subsystem has no entries for the priorities of appearance and comfort. The engine is inherited.

Figure 6.11 provides an example of how the power plant manager may respond to an action item to further break down requirements within the higher-level power plant allocations. Note in Figure 6.11 the absence of appearance and comfort in the allocations. Further, we see we are settling into the Neon, Saturn, MX-3, 200-SX, 240-SL, Civic, and Paseo power plant class.

The process continues under the leadership of the PDT team leader. Each level of the product WBS is addressed in a similar manner in more and more detail. The actual work is done by those most knowledgeable—subsystem managers, element managers, logistics personnel, test personnel, and so forth. As the process continues and hard requirements are understood, appropriate PDT members receive assignments to begin the generation of draft sections of the product requirements document. Alternative concepts often arise in filling out the structured matrices. When this occurs, the need for options analysis, or design trade-off studies, logically presents itself, and additional action items are generated for PDT members.

System Specifications

The system, or product, specification document (SSD) delineates "how" each function is to be realized. Functions are partitioned into hardware, software, and procedural functions. Detailed system specifications stating "how" each requirement is to be met for performance, physical characteristics, availability, environment, integrated logistic support, testing, and design and construction are provided. Functional area, or subsystem, requirements for each of the above items are also detailed.

An outline for the system specification document (SSD) is given in Table 6.2. While the structure is very similar to the FRD, the content provides significantly more detail. The important major difference between the two documents is that the FRD basically states "what" the system must do and the SSD states specifically "how" the system must meet the functional requirements. As with the FRD, the Table 6.2 outline is presented as a guide to assist the SSD developer in considering all pertinent factors. For example, the FRD might state a requirement as: "The system shall provide for storage of all institutional personnel records over the period of the system life." The SSD would address this requirement by including functional requirements for the size of personnel records, the number of records (allowing for

Automobile

PRIORITIES	Engine	Carburetion	Distributor	Exhaust
		POWER PLANT SUBSYSTEM		
SCHEDULE	X1 mos.	X2 mos.	X# mos.	--- ----- ----
COST	Y1 $ MTBF MTTR 28/33mpg Regular	Y2 $ MTBF MTTR	Y3 $ MTBF MTTR	----- --- ----- --- - --- ----- --- -----
PERFORMANCE	110 HP 4 cyl RPM 2.0 liters	Flow Mix	RPM	----- --- -----

(Neon, Saturn, MX-3,
200-SX, 240-SL, Civic, Paseo ---)

Note: **What's are beginning to
become How's**

Figure 6.11. Requirements decomposition process (Con't).

▌ TABLE 6.2 *System Specifications Document Outline*

1.0	Scope	
2.0	Applicable Documents	
3.0	Requirements	
	3.1	System Definition
		3.1.1 System Function
		3.1.2 Mission Descriptions
		3.1.3 System Diagrams
		3.1.4 System Interfaces
		3.1.5 Operational and Organizational Concepts
	3.2	System Characteristics
		3.2.1 Performance Characteristics
		3.2.2 Physical Characteristics
		3.2.3 Reliability
		3.2.4 Maintainability
		3.2.5 Availability
		3.2.6 Environmental
	3.3	Design and Construction
		3.3.7 Security
		3.3.8 Safety
		3.3.9 Human Engineering
	3.4	Documentation
	3.5	Logistics
		3.5.1 Facilities
		3.5.2 Transportation and Handling
		3.5.3 Personnel and Training
	3.6	Product Priorities
	3.7	Functional Area Characteristics
4.0	Constraints	
5.0	System Test Requirements	
6.0	Appendices	

growth during the product lifetime) and the response time for the retrieval of records. These are used to develop a specification for "how" the storage shall be provided. For example: "The system shall provide six megabytes of on-board RAM and 700 megabytes of disk storage to accommodate institutional personnel records over the product lifetime." A separate specification then states, "The system lifetime shall be 5 years." Note: each sentence is a single specification.

Similarly, interface flow discussed in Section 3.1 of the FRD is stated in English terms using N^2 diagrams, while the SSD discussion is in terms of specific formats to be used, the timing, and so on. In the same manner, the RAM section in the FRD covers the top-level system requirements. The SSD in turn responds to these requirements with a complete specification for the allocation of availability and logistics support mechanism across subsystems to achieve the product level requirements.

A request for proposal (RFP) should generally leave as much design freedom as possible for the prospective vendors. The FRD level of detail is thus preferred for an RFP with an appropriate constraint section. A request for quotation (RFQ) is a request for quotes on a specific detailed configuration or piece of equipment. RFQs are generally written using the SSD level of detail. As a rule, when an RFQ is issued, an FRD has already been generated and baselined to the critical design review.

Depending on the size and complexity of the system under development, subsystem specifications may be included in Section 3.7 of the SSD or may warrant their own documents. A separate subsystem specification document would typically be written if an RFQ for a subsystem were to be issued. The form and format follows the SSD specification with the system now replaced by the subsystem. Enough information is included at the system description level, however, to give the prospective vendors a clear understanding of the subsystems' role in the overall system. In this case, it may be acceptable to repeat all, or part(s), of Section 3.1 of the FRD in the subsystem specification. In general, however, simple repetition of document content in lower-level documents should be avoided to minimize the complexity of incorporating changes at a later time. The chore of updating documentation in a complete and consistent fashion can be greatly alleviated through the use of computer-based documentation aids.

All requirements stated in design documentation, subsequent to the FRD, must be traceable to their parent document in both forward and backward directions. This is the process of verification, which is an integral part of configuration management.

Software Requirements/Specifications

The content of the software requirements document (SRD) is similar in level of detail to the FRD. It addresses "what" the software must do in measurable terms. The specific requirements covered in the SRD are defined by the system specification document (SSD) where the meeting of functional requirements is allocated to hardware and software.

Numerous standards exist for the format of software requirements and specifications documents. The outline given in Table 6.3 for the SRD is a simplified composite of the software engineering literature and is intended to serve as a practical guide.

▐ **TABLE 6.3** *Software Requirements Document Outline*

1.0	Introduction	
	1.1	Purpose
	1.2	Applicable Documents
	1.3	Definitions/Acronyms
2.0	General Description	
	2.1	Problem Definition
	2.2	Data Flow
	2.3	Information Content
	2.4	Information Structures
	2.5	Design Constraints
3.0	Functional Partitioning	
	3.1	Program A
	3.1.1	Process Narrative
	3.1.2	Performance Requirements
	3.2	Program B
	.	. . . . . .
	.	. . . .
	.	
	3.X	Program X
4.0	Test Requirements	
5.0	Appendices	

Section 2 of Table 6.3 provides a top-level narrative description of the overall software problem with supporting top-level data diagrams and structured English. The description includes information content and structures, as well as database structures. Also implicit in the data flow description is a complete narrative description of software system interfaces. Design constraints, such as restrictions imposed by the target environment, may also be included in this section.

Section 3.0 of the SRD partitions the software in a fashion similar to the original system partitioning that takes place in the FRD. A proven method for software partitioning is the use of structured analysis techniques with data flow diagrams. Section 3.0 partitions the overall software problem into programs which become Computer Program Configuration Items (CPCIs). Narrative descriptions for each functional program and performance requirements are supported by lower level data flow diagrams. All data flow diagrams must include consistent interface descriptions and terminology where interfaces between CPCIs exist. All performance requirements must be measurable.

Section 4.0 of Table 6.3 addresses all software test requirements. These must account for all performance requirements. Test bounds, types, and classes are defined here along with anticipated results. During the actual day-to-day work of developing software requirements, it is expedient to use a software version of the traditional engineering notebook. The program development folder (PDF) provides a formal mechanism for the orderly assemblage and documentation of requirements, detailed design, and testing of software programs.

A major advantage of PDFs, or their equivalent, is that they give clear definition of tasks to be accomplished, and a clear record of progress toward that accomplishment, through the critical steps of program design, code, and test. For this reason the PDF can be a valuable aid to visibility for technical management. The PDF can also be of value to individual programmers by fostering consistent communication with other programmers, and by providing a basis for design documentation. The content of a typical PDF document is given in Table 6.4. A cover sheet identifies the programmer, the program covered by the PDF, and the PDF custodian. The custodian of the given PDF is normally the programmer or the programmers' immediate supervisor. The cover sheet also lists the contents by section and provides space for a due date and date completed for each section.

The requirements section (1.0 of Table 6.4) should give a top-level English description of the requirement(s) to be met by the program. It should also reference the relevant requirement(s) documented at a higher level. If the program requirements are derived in more detail

█ TABLE 6.4 *Program Development Folder Contents*

Section	Content
1.0	Requirements
2.0	Design Description
3.0	Program Source Code
4.0	Program Test Plans
5.0	Test Case Results
6.0	Deviation Log
7.0	Notes/Appendices

than the higher-level requirement, then that detail should be included in this section.

The program design is documented in section 2.0 using a structured approach adopted by the product development team (PDT). The design description contains the current detailed design for the program, which conforms to the established approach to structured design. The description is a living document during the design process. When concluded, it should be suitable for direct inclusion in the software design specification. A design walkthrough is normally held at the completion of the design description prior to program coding. The program source code, which is an extension of the design, is included in section 3.0 in the form of the current source code listing. The due date for this corresponds to the first error-free compilation of the program.

The program test plans section of the folder (4.0) contains a description of the testing approach to be used. This section identifies the required software drivers and/or tools to be used, test inputs, and expected outputs. Tests are planned to meet the requirements specified in section 1.0 of the folder and are traceable to higher-level requirements. In addition to functional testing, appropriate tests should be devised for error handling, range values, and logical path analysis.

Section 5.0 of the PDF contains a description of test results, successful or unsuccessful. Section 6.0 contains a record of any revisions to requirements, design, test plans, drivers, or tools required to achieve successful implementation of the relevant software program. Section 7.0, Notes/Appendices, includes any additional information necessary to clarify any and all aspects of the program implementation for any independent reviewer.

The PDT leader imposes rigor in the use and content of the PDF in direct proportion to his or her confidence that program implementation will be successful. The lower the level of confidence, the greater the need for details. Greater detail enhances the visibility, that is, understanding of the entire PDT. Typical reasons for an increase in the need for visibility are the use of large programming staffs, programming teams that are organizationally, or physically, separated, and issues related to technical feasibility. The use of a PDF (or an equivalent disciplined approach) can provide an important management tool for both the software PDT leader and the PDT leader. The PDF concept can also be used at lower levels of program design including computer program components (CPCs), and even modules, as required.

When visibility is required, the major advantages of the PDF are

- There is a clear and consistent understanding of what specific individuals are accountable for.
- The PDF provides management visibility at any point in time independent of reviews and walkthroughs.
- Design and test documentation is developed as it takes place.
- The PDF provides a mechanism for traceability.
- Communication among programmers (interface understanding) is enhanced.
- The PDF provides a mechanism for recording change as it occurs.

Software Design Document

The PDF may be sufficient as a software design document in many instances. A more formal alternative is to develop a software design document (SDD) in which detailed specifications are documented. An outline for a software design document is given in Table 6.5. The document is constructed in parallel with the design process, making successive use of the tools of structured analysis followed by structured design.

The design description (Section 2.0) makes use of data flow diagrams that are products of the preliminary software design activity. Program structures and program interfaces are reviewed and refined. This section includes complete descriptions of data types, files, and database structures and assignments of global data.

The functional description section (3.0) emerges as the detailed design proceeds. Programs are decomposed into computer program

❚┇ TABLE 6.5 *Software Design Document Outline*

1.0	Introduction	
	1.1	Purpose
	1.2	Applicable Documents
	1.3	Definitions/Acronyms
2.0	Design Description	
	2.1	Program Structures
	2.2	Data Description
	2.3	File Descriptions
	2.4	Data Bases
	2.5	Design Constraints
3.0	Functional Partitioning	
	3.1	Program A
		3.1.1 CPC 1
		3.1.1.1 Process Narrative
		3.1.1.2 Design Language Description
		3.1.1.3 Data Organization
		3.1.1.4 Test Requirements
		3.1.2 CPC 2
		_____
	3.2	Program B
		___
	3.X	Program X
4.0	Test Requirements	
5.0	PDL Representation	
6.0	Source Code Listing	
7.0	Appendices	

configuration items (CPCIs), computer program components (CPCs), modules, and procedures. CPC definitions begin with a narrative description, followed by successively detailed data flow diagrams at the CPC and module levels, and are finally represented using a program design language (PDL). Test requirements for each CPC are included with details at the module level as required.

Section 4.0 refines test procedures at the program level. Section 5.0 assembles a complete PDL representation of the software at a single location. This representation should ideally begin with a single page PDL description of the software under development. The next level may use three to five pages, and so on. PDL descriptions are organized in successive levels of detail around the software decomposition of CPCIs, CPCs, and modules.

When the system acceptance testing is completed, a source code listing is finally included as Section 6.0. This step ensures that a complete design description of the software is provided as a part of the software design document.

When constructed properly, the software design document should provide a complete and highly structured documentation set such that readers unfamiliar with the product can clearly find their way from a top-level understanding of product goals to the lowest levels of detailed code implementation. The software design document should be completed through section 5.0 in support of the critical design review, which should take place prior to the generation of source code.

Design Trade-Off Methodologies

Throughout product development, technical management is constantly challenged to acquire and maintain visibility. An important part of this understanding is gaining confidence that requirements and design concepts are realistic. Trade-off studies may be conducted throughout Phase C from initial consideration of architectural and technological options to detailed parts selection studies. Trade-off analysis is emphasized in Figure 6.1 because it commonly takes place to support early efforts to verify the technical feasibility of product functional requirements and in preparation for the system requirements review.

The basic components of a trade-off analysis consist of

1. definition of objectives
2. documentation of the approach
3. analysis
4. evaluation of alternatives

The definition of the objectives should be crisply stated in one or two sentences. The objectives must be measurable in order to provide a sound basis for decision and a well-defined end point. They should be structured to answer a single question. The measurable aspects are based on the ability of the proposed options to meet functional requirements and/or specifications. The study approach should be discussed by the PDT and documented. Approaches include analysis, modeling, breadboarding, brassboarding, building mules (automobile prototypes), prototyping, costing, risk analysis, and so forth. The analysis is carried out as an action item of the PDT. Alternatives should be evaluated using product priorities and the product WBS, as we shall see in the Phoenix example below.

A primary responsibility of the PDT leader is to lead in the identification of meaningful trade-off alternatives and to methodically guide the required studies to a productive and timely conclusion. The definition and execution of trade-off studies is a principal activity of the PDT under the leadership of the PDT leader. The nature of trade-off studies may vary from architectural and system performance analyses early in the process, to evaluation of specific technology alternatives during detailed design to the weighing of proposed design changes late in the overall process. In conducting such studies, it is highly desirable to employ techniques that are as objective as possible and provide a consistent basis for decision making throughout the development process.

Suppose we wish to conduct a design trade-off for our Phoenix automobile. The question is: Should we use a four-cylinder or six-cylinder engine? Table 6.6 shows a starting point for the process. Major subsystems are listed across the top of the page and priorities listed on the left. The PDT uses the diagram to brainstorm possible impacts of the design trade on each subsystem as it relates to each of the priorities. At this point, our mind-set is simply to consider whether

▌ TABLE 6.6	*Using Priorities in Trade-Off Analysis: Phoenix Automobile: 6-Cylinder VS. 4-Cylinder Engine*				
	SUBSYSTEMS				
Priorities	**Power Train**	**Chassis**	**Body**	**Electrical**	**Assembly**
Schedule	1	2	3	4	5
Cost	6,11	7	8	9	10
Appearance					
Comfort		12	13		
Performance	14	15			

a potential impact exists. We will develop statements of actual impacts later. For the moment, we are just concentrating on whether there is a potential impact. For example, there may well be an impact on schedule. We don't know yet, so we enter a number for each subsystem 1 through 5. That means we will want to develop some kind of specific action item for each of the subsystems, or at least one that covers all five. A similar situation relates to cost indicated by entries 6 through 10. We note there are at least two cost considerations for the trade-off related to the power train, one is the cost of the four- versus six-cylinder engine. The other is the cost associated with maintenance—reliability and serviceability (entry 11). Hence, there is a call for two potential action items in the matrix at this point. There are no potential impacts related to appearance of the Phoenix because you can't see the engine from the outside of the vehicle. The six-cylinder engine might possibly require a larger chassis. If this is the case, the body may also be affected. These factors have a potential impact on comfort (entries 12 and 13). Finally, the differences in performance need to be considered, as well as the potential impact of each engine on the chassis (entries 14 and 15).

Next we begin to define each issue in terms of top-level action items as shown in Table 6.7. We then translate these top-level considerations into more specific trade-off study needs that are measurable as shown in Table 6.8. Action items are assigned to appropriate PDT personnel, with schedules for completion and reporting back to the team. Table 6.9 continues the example with statements of findings for each action item. Table 6.10 then provides a summary of the outcomes. The first thing we note in this example is that the trade-off has no impact on schedule or on vehicle appearance. These are the first- and third-highest priorities of design (Table 6.6). The second-highest priority, cost, favors the four-cylinder engine. The findings on comfort and performance are mixed. There is no impact on interior comfort or trunk space. The only advantage of the six-cylinder is a smoother ride and improved braking. But these priorities are ranked fourth and fifth. The team selects the four-cylinder option solely on the basis of ranking of cost versus the more marginal differences in comfort and performance. Further, since the same chassis can be used, it would be possible to offer a six-cylinder version sometime in the future. The four-cylinder decision fits with all our priorities and is consistent with the mission statement.

There are some important points to be made regarding the process outlined in the example. First, we are confident that we have covered all aspects of the product itself because as a team we considered each item of the product WBS in the original matrix. Further, the

▮ TABLE 6.7 *Using Priorities in Trade-Off Analysis: Verbalize Potential Action Items*

Notes	Top-Level Action
1–5	Determine schedule impacts on components of each subsystem.
6–10	Determine cost impacts on components of each subsystem.
11	Develop estimates for power train miles per gallon, reliability, and maintainability.
12	Determine any impacts on chassis size and on suspension impacts on ride.
13	Determine impacts on interior comfort vs. chassis size.
14	Determine impacts on horsepower, 0–60 MPH, and maximum speed.
15	Determine impacts on braking and steering.

▮ TABLE 6.8 *Using Priorities in Trade-Off Analysis: Develop Specific Action Items*

Notes	Action Item
1–5	Identify longest lead time components for each subsystem, for each alternative and overall schedule impacts.
6–10	Estimate specific unit cost impacts for design and production at component level and sum to product level for each subsystem. Identify cost impacts for assembly.
11	Develop MTBF and MTTR data for power train components and for power train. Estimate miles per gallon.
12	Quantize chassis size effect on ride (vibration, motion, etc.).
13	Rate interior comfort vs. platform size using existing models.
14	Quantize 0–60 MPH time, maximum speed, and HP in terms of percentage differences between two engines.
15	Quantize braking and steering in terms of percentage differences.

❚❘ **TABLE 6.9** *Using Priorities in Trade-Off Analysis: Action Item Findings*

Notes	Action Item
1–5	No impact for power train, chassis, body, electrical, assembly.
6–10	8% increase in retail price, negligible assembly impact.
11	6-cylinder MTBF 4% higher, MTTR 10% higher, MPG 33% lower.
12	Use same chassis, 6-cylinder smoother, less noise, less vibration.
13	No interior comfort impacts, no trunk size impacts.
14	6-cylinder 0–60 8% faster, HP 30% higher.
15	6-cylinder braking 10% faster, same turning radius.

❚❘ **TABLE 6.10** *Using Priorities in Trade-Off Analysis: Example Outcome*

Priorities	Outcome Summary
Schedule	No impact.
Cost	Favors 4 cylinders—retail and maintenance.
Appearance	No impact on body.
Comfort	No impact on comfort, trunk space, same chassis. Favors 6 cylinders for ride, noise, acceleration.
Performance	Favors 6-cylinders for braking. Can meet 0–60 MPH in 10 sec with 4-cylinder.

relative importance of each action item finding was determined by playing them against our established product development priorities. In a sense, the process is deceptively simple. It is interesting that in the vast majority of cases, this process will clearly lead to selection of the best alternative.

In responding to action items related to trade studies, some type of physical or analytical modeling is often required. Modeling should be used at any point in the process where confidence in performance issues is at all questionable. The purpose of modeling is to mimic the appearance, or behavior, of a system, or subsystem, so that an improved understanding of unknown attributes can be acquired without the time and expense of building and testing the final product.

Modeling

We are all familiar with concepts of early and rapid prototyping, breadboarding, mock-ups, wind tunnels, and so on, as well as operational research tools such as simulation and analytic modeling. In one sense, these are all distinct disciplines and, indeed, whole careers have been devoted to their subtleties. From the PDT leader's viewpoint, however, the use of each of these tools has a common motivation—the reduction of uncertainty through the study of physical and/or mathematical behavioral representations.

The need to reduce uncertainty can arise at a number of points in the product development process, from determination of requirements to the response to specific trade-off action items, to validation of producability of the final baseline product. Each modeling technique exhibits its own value with respect to the degree to which it can reduce uncertainty as a function of the specific issue at hand. The PDT leader should have a clear perspective as to the purpose for undertaking a particular modeling exercise.

There are two fundamental purposes for undertaking modeling. The first is to assist in defining requirements themselves, and the second is to verify an implementation concept designed to meet these previously defined requirements. This may seem like a moot difference, but technical management is commonly subject to criticism when requirements "change," particularly as the configuration management structure becomes more formal and change becomes more difficult. It is not unusual for a modeling exercise in support of a design process to turn up surprises that result in the need to alter requirements. It is often advantageous to conduct modeling early in the process in order to verify that specific 'more difficult' requirements can be met. If there are any doubts, it is best to recognize this situation and schedule the work at the requirements definition phase. This reduces the risk of "changes" later in the process. The choice of modeling techniques and the timing of their use is inherently related to the product development paradigm in use.

It is also important that management understand the strategies employed and their rationale. For example, early prototyping is commonly used to understand user requirements. In this application, the modeling work is not intrinsically a top-down detailed design process. The prototype is rather brought together quickly with emphasis on

ⓘ TABLE 6.11		*Framework for Modeling in the Synthesis of Requirements and Design*
Class	**Type**	**Examples**
Static	Physical	Mock-ups, positioning, static loading, etc.
	Math	Queueing, reliability, availability, maintainability, steady state analysis, etc.
Dynamic	Physical	Breadboards, brassboards, early prototyping, mules, production prototyping, and analogues such as wind tunnels, ship hull models, biological models, etc.
	Math	Simulations such as GPSS, SIMSCRIPT, differential and integral models, etc.

resolving specific issues. With luck, much of the prototype development may be usable in the actual detailed design, which is to come later, but this is not usually the case, *nor should it even be counted on.* Management often misunderstands this and asks, "If you have built a prototype that works, why do you need further design time?" Thus, it is important to understand, and to make known, the specific goals and expected products of modeling at any particular stage. Such steps to reduce uncertainty should be a planned and well-advertised part of all schedules.

Table 6.11 breaks down the approaches to modeling by class and type, with examples. Models are divided into static and dynamic classes. Each of these classes can, in turn, be divided into physical and mathematical types. The following discussion focuses on the roles of these modeling tools in the synthesis of requirements and design, and provides general guidelines for timing and conditions for their use from the PDT leader's viewpoint.

Static models are primarily concerned with steady state conditions. The first type of static model is physical. Examples of this type include mock-ups (such as cockpit, console, or workstation layouts), and the physical laying out of simulated equipments, cables, and the like, in order to mimic design and/or placement concepts. Mock-ups are generally used to bring more precise definition to requirements affecting human factors, safety, and physical fit issues. Static loading models on structures such as airplane wings, and the like, are employed to verify that a design concept and/or specific materials meet previously defined requirements. The second type of static model is mathematical. Static mathematical models include tools such as queueing analysis, availability, reliability and maintainability analysis, and any of a large class of analytic representations that predict steady state conditions. The disciplines of queueing analysis and system availability analysis are of extensive value to the PDT. The majority of complex systems built today are characterized by the use of computers involving contention for resources and stringent system availability requirements. In such systems, queueing analysis (or simulation when required) and availability analysis are the principal tools used to initially determine top-level system architectures.

Dynamic models are also divided into physical and mathematical types. Dynamic physical models (or physical analogs) include, for example, the construction of crude approximations, near replicas, use of scaled-down models, and experimentation with systems that exhibit similar physical characteristics. Familiar scale models include ship hulls in water tanks and aircraft models in wind tunnels, scaled-down automobile designs, and so forth. Biological models are used as analogs of human systems; for example, the hind legs of a cat represents the closest analog to the musculature of the human lower extremities.

Dynamic mathematical models include transaction-oriented simulation, or discrete, simulation tools, such as SIMSCRIPT and GPSS, as well as the use of differential and integral calculus analogs implemented on digital and, less frequently, analog computers (if you can find one these days). Typical uses in these arenas include analysis and design of computer systems, communications systems, fire-control systems, engine performance, marketing systems, vehicle vibration studies, economic systems, physiological systems—virtually any continuous system.

The classifications and examples presented here are, of course, not exhaustive. The structure presented is intended as a conceptual framework that can easily be amended.

It is useful, at this point, to discuss in more detail some of the techniques used in dynamic physical models. Tables 6.12 through 6.15 summarize the principal properties of breadboards, brassboards, early prototypes, and prototypes. These terms are familiar to most engineers. However, it is worthwhile to characterize them with respect to the timing and rationale for their use in support of the PDT design process.

Breadboards (Table 6.12) provide a quick and dirty look at concept verification. Electronic breadboards are usually a stringy mess and do not provide a good environment for tolerance-testing of factors

▯: TABLE 6.12   *Characteristics of Breadboards*

Characterized by:	A stringy mess
Confidence level:	Low
Rationale:	Testing of new ideas
Be sure to:	Keep a good engineering notebook

▯: TABLE 6.13   *Characteristics of Brassboards*

Characterized by:	A tidy semikluge
Confidence level:	Medium
Rationale:	Test out a few unknowns before committing
Be sure to:	Keep a good engineering notebook and thoroughly test

▯: TABLE 6.14   *Characteristics of Early Prototypes*

Characterized by:	Near replica of working environment
Confidence level:	Medium
Rationale:	Definition of unknown requirements
Be sure to:	Keep a good engineering notebook and minimize configuration control

▯: TABLE 6.15   *Characteristics of Prototypes*

Characterized by:	Replicas of serial number one
Confidence level:	High
Rationale:	Build and test serial number one before production
Be sure to:	Duplicate production model and thoroughly test

such as the effects of capacitance, cross-talk, and the like in a finished version. Breadboards provide an initial test bed only. When they work, a more tidy representation closer to the final product should always follow.

Brassboards (Table 6.13) generally represent a more serious attempt to mimic the behavior of the final product, particularly in terms of component placing, connector lengths, and packaging. Portions of brassboards that involve high-risk areas may be constructed as near replicas of the envisioned final product and may be of sufficient integrity to undergo considerable testing.

A distinction is made between the terms "early prototype" (Table 6.14) and "prototype" (Table 6.15). Early prototypes are near replicas of the working environment. For example, if an electronic breadboard proved successful, an early prototype may then be constructed to resolve any potential capacitance problems. Prototypes are virtual replicas of serial number one and are built primarily for extensive testing and for ensuring producability.

Breadboards, brassboards, and prototypes are typically built to support design trade-off studies or to verify the capability of more mature designs to meet previously established requirements. Early prototypes are quite different in that they are usually used to assist in the definition of requirements, or in some instances, trade-off analyses. Their design is characterized by providing a near replica of user interaction even though the location might be quite different. A further principal characteristic is that the development environment is ideally capable of rapidly accommodating potential design changes in response to user feedback.

Early, or rapid, prototyping finds its greatest value when the user is unable to communicate the exact requirements. But the user is able to recognize, select, or provide guidance about desired design features when confronted with appropriate alternatives. Such conditions commonly exist, for example, when building the user interface for a new military command and control system, for a novel management information system, or for the support of market research for consumer products.

Rapid or early prototyping can be of value when requirements are unknown or fuzzy, but not always. In some cases, exact requirements can not be articulated because no one inherently knows what they are nor can they be recognized in any detail when shown. This is the very real case, for example, with the space station whose mission will extend beyond a time frame that enables anyone to foresee it's exact usage. In cases of this sort, the systems engineering process should be structured around the rapid development concept.

For the PDT leader, the choice of model implementation is predominantly driven by the level of confidence associated with the inquiry at hand—that is, the depth of inquiry required and an understanding of whether the issue has to do with establishing requirements or with design issues. Model structures are often improperly chosen. Engineers who spend their lives in rapid prototyping laboratories will naturally tend to favor rapid prototyping to resolve as many issues as possible. I have also seen simulation experts spend a great deal of time and money in order to resolve system performance issues that could have been determined in one-tenth to one-hundredth the time through the proper application of simple queueing theory. I was once asked by a colleague to review the results of a simulation effort that had taken some seven months to complete. It was a classic queueing problem. In the next day and a half I was able to obtain the same answers by using a combination of analytic queueing models. He was a simulation guy. It never occurred to him to use an analytic model. When I explained how I got the same answer, he was not happy.

In summary, it is best to clearly consider the following points when electing to employ modeling techniques:

1. Discuss the need and strategy for modeling with the PDT. Reach a consensus on the right approach. Call on experts if you have any doubt. Derive and write down a succinct statement that addresses the specific question to be answered in measurable terms. The goal must be limited and well defined. One or two sentences for inclusion in the minutes should be sufficient.

2. Construct a realistic schedule to achieve the acquisition of resources, as well as execution of the analysis and of required iterations. The schedule must support the project schedule.

3. Assign lead responsibility for resolution of the question to a single individual.

4. Be sure that all required data input and/or drivers for the model can be accurately acquired in a timely manner. These are often overlooked.

5. Limit the activity to addressing the question at hand.

6. Inform management of your plans.

Interface Development and Documentation

A fundamental responsibility of the PDT leader is to maintain control over the definition, development, design, construction, and test of system, segment, element, and subsystem interfaces. This point was emphasized through the soccer ball analogy shown in Figure 3.1 in Part 2. The outline of the ball represents the system boundary, which receives inputs from, and delivers outputs to, the outside world. The seams in the soccer ball depict the boundaries between internal decompositions within the system. These internal decompositions (segments, elements, subsystems, and so on) can be viewed as systems within the overall system. The PDT leader is responsible for the definition and control of the flow across both system and internal system boundaries. The external and internal boundaries are the lines upon which the system engineer must maintain coordination to meet system-level requirements.

The internal system cognizant engineers (COGEs) are deeply enmeshed in the details of internal system design and fabrication. The PDT leader typically follows this detail only to the extent necessary to maintain the integrity of the interfaces. Two widely used methods for the organization and documentation of interfaces are the N^2 diagram and the open system interconnect (OSI) models.

The N^2 Diagram

The N^2 diagram provides a structured method for the initial definition of interfaces and their successive detailed definition at lower levels[1]. The general structure of the N^2 diagram is shown in Figure 6.12. Subsystems are positioned on the diagonal of the diagram. Subsystems A through F are depicted as shaded in the figure. Subsystem outputs are displayed as horizontal lines, and subsystem inputs are displayed as vertical lines. The absence of a box at the appropriate location of the matrix means that there is no direct interface between the designated subsystems.

In the example of Figure 6.12, subsystem A has an output that becomes an input to subsystem C, denoted by the box with text "from A to C." In fact, subsystems A and B both have inputs to and outputs from subsystem C. But note that subsystem A has no direct inputs to or direct outputs from subsystem B. Similarly, subsystem D has inputs to subsystem C but has no direct inputs from subsystem C. Subsystem F has outputs that become inputs for all other internal systems.

The interface definition process begins at the top level and proceeds to lower levels. Iteration between levels takes place as more insight is gained and design decisions are made. Initially the interfaces are described in English, with more detail added at each successive level. Interface particulars are eventually expressed in precise detail using tools such as program design languages, circuit data sheets, or engineering drawings.

270

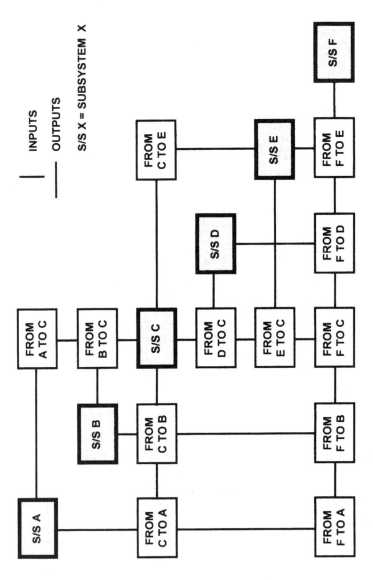

INPUTS

OUTPUTS

S/S X = SUBSYSTEM X

Figure 6.12. The N² interface diagram.

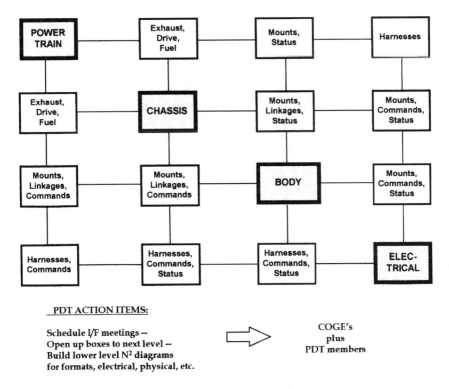

PDT ACTION ITEMS:

Schedule I/F meetings --
Open up boxes to next level --
Build lower level N² diagrams
for formats, electrical, physical, etc.

COGE's
plus
PDT members

Figure 6.13. Automobile N² diagram example.

By way of example, consider our old friend, the Phoenix automobile. Each major subsystem of the automobile is shown on the diagonal of Figure 6.13. The definition process begins by entering a top-level description of each interface where appropriate. For example, the power train has exhaust, drive, and fuel interfaces with the chassis, mount, and status inputs to the body, and harness interfaces with the electrical subsystem. Body outputs to the power train include mountings, linkages, and commands. We begin at a very top level.

Interface definition meetings are scheduled by the PDT. These interface definition meetings are working meetings and take place outside the PDT meetings. For example, the power train and chassis manager receive an action item to further describe their interfaces to lower and lower levels of detail. The detail continues until a specific design is attained. Every detail eventually becomes very specific. For example, the body has the interior, the interior has a dashboard, the dashboard has instruments, and the instruments are lit from behind. Let's say the instruments are to be lit by light bulbs. Each bulb has a socket with wires attached to it. Who is responsible for what? A reasonable answer is that the electrical subsystem supplies the wires from a harness, the bulb socket, and the bulbs. The body subsystem would then provide mounts for the sockets behind the instruments. In this manner, exact interfaces between subsystems are defined and are reflected in subsystem requirements, specifications, drawings, and so forth, as the detailed design emerges. Formats for status messages are also agreed to and standardized by the process. Interfaces between components and parts within subsystems are further defined. Figure 6.14 shows an N^2 diagram for a steering system that is in the working stage. The diagram reflects the early thinking among those responsible for the steering system. For example, will the electrical system require status information from the mechanical system? The steering team will need to decide that and pinpoint what that status is to be, exactly how it shall be provided, and who will provide each side of the interface. The question marks denote a need for further definition.

A distinct advantage of this methodology is that interfaces are defined from the top down. This provides a logical and clear structure for the delineation of design issues at each subsequent level. The formulation represents a structured methodology, and hence opportunity, for the PDT leader to direct the unfolding of needed detail in a logical fashion. The process is carried forward with more and more detail involving negotiation and issue formulation by the COGEs and the PDT leader. Throughout this process, the PDT is the focal point for the identification of pertinent issues and the assignment of action items to explore alternatives outside the PDT meetings. (Remember, PDT meetings are not working meetings.) The PDT leader leads this identification and assignment effort with particular care given to addressing all issues at the proper level of definition.

The OSI Software Interface Model

Another useful means of visualizing the levels of detail when dealing with interfaces is provided by the open system interconnect (OSI) model.[2] The OSI model was initially developed by Subcommittee 16 of the International Organization for Standardization. The goal of the OSI model is to bring universal compatibility between computers, computerized equipment, and communications networks from vendor to vendor and ultimately from nation to nation. The term "open" refers to the concept that any system developed with the OSI model would be open to all others that used the model.

The OSI model consists of seven layers. These are summarized in Table 6.16. Examples of the functions at each layer are given to assist in determining the use of each layer for a given implementation. Efforts

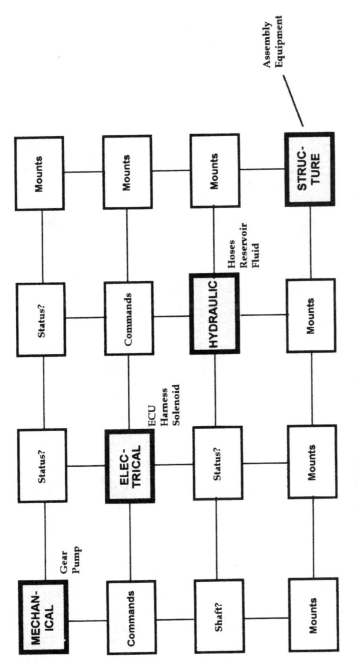

Figure 6.14. Power steering N² diagram example.

▌ TABLE 6.16   *Open System Interconnect Model Layers*

Layer	Function	Examples
Application	Direct support of user, system management	Batch, data entry, DBM, RJE, reservations, word processing, banking, process control, etc.
Presentation	Prepares data for direct manipulation by users	Display formats, code conversion, language, terminal emulation, data exchange, etc.
Session	Manages end-user dialogue and resources	Log on, user verify, equip. allocation, token management, etc.
Transport	End-to-end control for session path	Msg. blocking and assembly, mux, error control, sequencing, failure recovery, etc.
Network	Manages nodal routing	Addressing, routes over nodes, flow control, etc.
Data Link	Node-to-node protocol	Synchronization, error control, ACK, NAK, etc.
Physical	Electrical/Mechanical	1553, 232C, 499, etc.

are still underway to bring precise definitions to the upper levels. While this work is still in progress, the user can generally be viewed as interacting at the application level. Some products do not require all seven layers. For example, a computer-to-computer dedicated communications system may require the bottom three levels only, involving specification of the electrical connection, a suitable protocol, simple addressing, and error control. In this case, the user would view the ISO model at the network level. On the other hand, a telephone system involves simply the physical and network levels. That is, there is no node-to-node protocol, but there is a physical connection and addressing is present (nodal routing) through the dialing mechanism.

Figure 6.15 provides an example of the interpretation of the OSI paradigm as applied to the execution of built-in test equipment (BITE) in a system. The built-in-test (BIT) manager issues a command to initiate BIT at the application level. The command has an address that is noted as a network-level function. In this view, there are no functions at the presentation level for the issuing of the command. The command is issued to the physical layer using the protocol introduced at

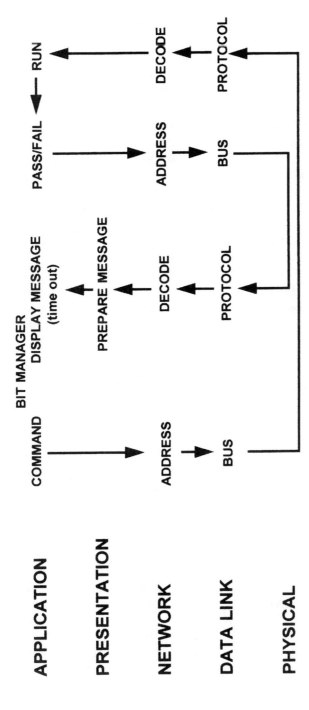

Figure 6.15.   Built-in-test (BIT) example using OSI Model.

the data link level. The command is received at all potential destinations and decoded at the network level. The addressee then initiates BIT and determines the pass/fail status at the application level. The status message is addressed at the network level and sent through the data link and physical layers to the host. After message decoding at the network level, a status message is prepared at the presentation level and the message is then displayed at the host application level.

The ISO model has value in that it provides a structured approach to issues associated with successive levels of design. As such, it represents another tool that can be useful as a guide in organizing a top-down approach in conjunction with the use of N^2 diagrams.

Interface Control Documents

Interface requirements and designs are commonly assembled in documents called interface control documents (ICDs). There are no widely accepted standards, military or other, for the format and content of ICDs. However, these are some useful guidelines.

There should be one ICD for each interface. This should take the form of one document, or a single chapter in a document, that covers the complete interface for each of the shaded boxes that appear in your N^2 diagram. Once generated the document is then signed off by at least two responsible cognizant engineers and the PDT leader. The ICD should be a single repository for interface requirements and the detailed interface design. This means that it is a living document throughout development up to the critical design review—the freeze point. It also means that the finished document contains the detailed specification of the finished product and thus has utility as a user's manual.

An outline for the ICD is given in Table 6.17. Section 3.1 furnishes a brief descriptive overview of the function of the interface. The level of description is similar to that provided for the top-level English descriptions at the higher levels of the N^2 diagram.

The interface requirements (3.2) are divided into software requirements and hardware requirements. The software requirements are for functions between ISO model layers, or for modules that interact within a single ISO level, above the ISO physical (hardware) layer. These requirements should include data, message types, rates, transfer control mechanisms, formats, content, accuracies, response times, error detection and recovery, protocols, and all other functions that may be defined from the ISO application layer through, to, and including the data link layer (see Table 6.16). The software interface description (3.2.1) should include a clear statement of the sequence of message/data interchange that transpires between the interfacing software entities for each type of information transfer.

☐ TABLE 6.17 *Interface Control Document Outline*

1.0	Introduction/Scope	
2.0	Applicable Documents	
3.0	Requirements	
	3.1	Functional Description
	3.2	Interface Requirements
		3.2.1 Software Requirements
		3.2.2 Hardware Requirements
	3.3	Environmental Requirements
	3.4	Test Requirements
4.0	Design	
	4.1	Software Design
	4.2	Hardware Design
5.0	Appendices	

The hardware requirements section (3.2.2) includes all physical, electronic, mechanical, optical, hydraulic/pneumatic, and so on, interfaces. These include such items as the ISO communications physical layer, radios, mechanical matings, and the like. The finished ICD should contain detailed information in a clear and appropriate form, such as mechanical drawings, interfacing circuit component schematics, pin assignments, electrical wave forms and timing, circuit data sheets, and so on.

Environmental requirements (3.3) include items such as electromagnetic compatibility, radiation, thermal, pressure, vibration, shock, and dynamic and static loading.

Section 4.0 documents the interface design. The appendices (5.0) should include additional information that may be required to clarify aspects of the interface, as well as a glossary of terms. The ICD outline given in Table 6.17 is not a hard standard. It is rather a guide that fits most cases. Modification to this structure may be adopted as circumstances require. The proposed form and format for the ICD is first produced by the PDT leader. It is then presented to the concurrent product development team for discussion, finalization, and concurrence.

Product Test Plans and Procedures

This section discusses the basic concepts of test plans and procedures. The concepts apply to all levels of testing, from parts and unit testing to complete product testing at the system level.

Formal testing strategies are delineated as test plans and procedures (TP&P). The test plan and procedures are formal documents derived from product level test requirements stated in the functional requirements document and in the software requirements document. The plan part of the TP&P document specifies how each specific test is to be implemented. The plan specifies the test verification approach, equipment to be used, types of personnel required, an overall schedule, anticipated results, the format for test report documentation and a clear traceability matrix for each functional requirement.

The plan for testing in support of product development is structured around four major conceptual test activities. These are unit testing, system integration and test (SI&T), system testing, and acceptance testing.

Testing that supports development should test every hardware and software functional requirement, including interface requirements, at some point over the range of tests from unit to full system testing. Acceptance testing may, or may not, involve a complete repetition of the developers' system tests. In all cases, test plans in support of system development should include plans for product acceptance testing.

There are four basic verification techniques that apply to all four major testing activities:

1. *Inspection*—basically the visual examinations of engineering drawings, software designs and/or listings, configurations, the final product, and so forth.

2. *Analysis*—use of simulation and/or analytic modeling, or other analyses, to verify performance limits and accuracies; used when direct testing is deemed to be too complex, time-consuming, or costly.

3. *Demonstration*—a verification method less stringent than detailed testing and used to illustrate top-level functional go/no-go capabilities.

4. *Test*—employed to verify requirements through the use of equipment, precise measurements, and detailed procedures.

Test planning is an important exercise. Experience has demonstrated that the following errors, or inadequacies directly related to testing, are the most predominant causes of unforseen problems. These problems can range from development schedule/cost impacts, to increased operational costs, to outright operational failure. Among the potential errors, the following are the most common:

1. *Inadequate low-level testing*—refers to the failure to conduct adequate testing of hardware parts, subassemblies, components, and software procedures, modules, and CPCs.

2. *Inadequate consideration of test impacts on design*—lack of sufficient test points with regard to strategic function and/or number. For example, this usually occurs with placement of too many parts on a single board or platform. This is an assembly and subassembly partitioning issue.

3. *Conservative environmental tests*—negligence in executing tests over the complete range of environmental extremes over a sufficient period of time.

4. *Deviation from test plans & procedures*—this most typically happens when schedule and cost limitations are encountered due to unforseen extensions in regression testing. Programmatic pressure results in the subordination of adequate retesting at system levels. Be aware that the staircase and staircase with feedback paradigms are not only most susceptible to this, but that deviation is an extremely common occurrence attending the use of these models.

5. *Inadequate early planning for test logistics support*—this oversight involves the discovery that additional test equipment, or other testing resources, such as software drivers and stubs, are required in addition to those originally planned in the TP&Ps. Other logistics oversights include inadequate planning for test equipment failures and the logistics of pre- and post-shipment accommodation details.

6. *Insufficient aggressiveness in testing*—primarily motivated by the natural overwhelming desire to pass, which commonly supplants the more intelligent mind-set that the finding of errors, in the long run, is a positive event.

The test plan must be carried to sufficient detail to gain confidence that a detailed design of test procedures can be realized with acceptable risk. Test procedures draw upon and expand the test plans developed during the preliminary design phase and are added to the test plans and procedures document.

The procedure part of the TP&P specifies in detail the exact sequencing of events, specific daily schedules, required instrumentation, configurations, support equipment, personnel assignments, complete traceability mechanisms for validation of requirements, and expected detailed results with appropriate reporting forms. The report forms include a section for test anomaly reporting.

For example, the detailed procedures for radio propagation contour measurements may include driving an instrumented vehicle from point A to point B on a specific day and measuring signal strength at input to the first intermediate frequency stage of a receiver at specified time intervals. A software test example might be the insertion of an exact test word in a specific register, execution of a particular code path to a breakpoint, and examination of a predicted bit pattern in a second register. Detailed test report forms include anomaly reporting and accommodation for retesting and reporting.

TP&P documents are routinely developed for unit testing, subsystem testing, integration testing, complete system testing, and for acceptance testing. They are also developed for unit testing at the hardware configuration item level and computer program configuration item levels. Test plans and procedures should be coordinated across all of these efforts because the success or failure of one has substantial impacts on all the others. It is also true that the ability to execute test procedures may easily have serious impacts on equipment and software designs. The following paragraphs provide further observations on the four major test activities.

Unit Testing

Unit, or subsystem, testing is a work item under the development of mission product items and is the responsibility of subsystem cognizant engineers (COGEs). In this scheme, fully tested and functioning subsystems are submitted for integration testing. While unit testing is primarily within the province of the COGE, there are still many system-level issues to be considered. The ability to test a unit has a strong impact on design, as does the ability to test at the SI&T and system levels. These testing strategies should be coordinated in order to make use of similar validation and verification methodologies as often as possible. The development of such coordinated test procedures is a system issue, hence very much a part of the PDT agenda.

The PDT leader should also be aware of the many productive testing strategies that can be carried out below the subsystem and program levels. These include complete or probabilistic parts testing upon receipt or fabrication, subassembly and component testing, and thorough testing of critical software algorithms, procedures, modules, and CPCs.

System Integration and Test

System integration and test (SI&T) consists of the methodic interconnection and testing of hardware configuration items and computer program configuration items in increasing numbers and in increasing complexity until the entire system is interconnected and its interfaces are completely tested. In complex systems, SI&T marks the first time in the development cycle that the PDT leader—and everyone else—is able to gain hard evidence of the probability of success. Other tools, such as prototyping, simulation, structured code, peer reviews, code walkthroughs, and so on, may have been employed prior to this point to enhance confidence in the integrity of the design and fidelity of the implementation. However, this is really the first time in the development process that the actual system initially comes together—and works or doesn't work. It is a time that can be fraught with disaster and stymied with much finger-pointing, particularly when multiple vendors, or organizations, are involved. It is also a time when the professionalism and cooperation of all parties involved can be seriously strained. One rule of thumb in developing schedules and procedures for SI&T is to take the best estimate you and your associates can make. When you are all comfortable with your guesses—multiply the time line by two (some say three!). The major reason SI&T can be such a lurking problem is that when a deficiency is found and corrected, the correction needs to be retested. A basic question, always confronted at this point, is whether to perform this retest in isolation—that is, test for the exact fix, or repeat integration testing to that point in order to ensure that the fix has not introduced other subtle discrepancies. The problem is most severe in software testing where such subtleties are often not apparent.

The repetition of integration testing is called regression testing. In its simplest form, regression testing requires the complete reset of the SI&T clock every time it is performed. While partial regression testing is less demanding, it can still be extremely difficult to anticipate the magnitude of retesting. It is also difficult at the project planning phase to determine the exact methods to be employed for retesting. This important issue is addressed in the construction of detailed test procedures for SI&T.

There are a number of meaningful steps that can be taken in system and subsystem design to guard against the spiraling chaos to which SI&T can easily degrade. When something goes wrong, the whole name of the game is fault isolation and rapid retest. In hardware, it is difficult to overestimate the importance of strategically located, functionally oriented test points. It is very important to instill this awareness in each designers' thought process, even to the point of conducting special internal reviews on the subject. If you can afford it, it is better to have too many test points than to wind up saying, "If I only had a test point—there!" Early, extensive, well-planned component, subassembly, and unit test prior to SI&T can also greatly reduce malfunctions. Identification of faults at the component level is considerably cheaper per fault than identification at the SI&T level.

There is also a distinct tradeoff between the size of subassemblies to be tested and the complexity (hence costs) of the tests themselves. The decision of what goes into a subassembly directly impacts the modularity of the design and the interface design. In some cases, testing requirements alone can dictate subsystem partitioning. There is no doubt that when dealing with highly complex systems, experienced test personnel are a significant asset to the PDT. The payoff is often more than evident in SI&T.

In software design, functional modularity and high cohesiveness are key features in augmenting ease in fault isolation. A major virtue of object-oriented programming lies in the relative ease of association of functional malfunctions with specific functional modules. However, one of the most effective methods of early fault detection in software—a method seldom practiced—is simple peer review of source code prior to compilation. Code and design walkthroughs, especially in areas where your confidence may be less than total, are also a must.

Extensive unplanned software regression testing is without question the single-biggest threat to a testing schedule. It has been termed the "black hole" of software development. When testing begins to drag out beyond the "planned" time, there is increasing pressure to get the job done. Tough decisions have to be made to minimize further delays. Complete regression testing can easily give way to partial regression testing, and next give way to simple verification of isolated fixes. The worst comes when the execution of TP&Ps are degraded or even abandoned. Under these pressures, configuration management can become a nightmare. Not far behind walks the credibility of the entire development effort. *Have smart test people on your PDT.*

System Test

There is a subtle distinction between SI&T and system-level testing. SI&T deals with the methodic, sequential assembly and testing of system and subsystem interfaces. System interfaces are, in fact, interfaces of specific internal systems to the external world, and hence may have been tested to some degree in SI&T. However, the emphasis of system testing is on system-level responses to system-level inputs—that is, the response across all internal systems that a particular test path may take.

System testing is carried out by the implementing organization prior to any customer involvement in testing. The purpose is to verify that all system-level requirements are met. These include all end-to-end information flow requirements, as well as environmental, performance, and system availability requirements. As with all testing, system test procedures are traceable to system test plans which in turn are traceable to each functional requirement.

When the implementation organization is satisfied with system-level testing, it is basically ready to begin customer acceptance testing. It is common to conduct acceptance testing at the user's site. System testing, therefore, is often divided into two distinct iterations. The first is prior to system breakdown and shipment to the vendor's site. The second occurs after shipment and installation at the user's location. Ideally, system-level tests at the vendor's site, prior to shipment, should take place using the *identical* configuration that will be used at the user's site. This includes such details as the same placement of equipment, use of cables, environment, and so on. Pre-ship and post-ship reviews are often scheduled to ensure the readiness and adequacy of these system-level tests.

Deviations from this course may arise if it is expedient to deliver parts of the system directly to the customer's location rather than to a central vendor site for system testing. In these cases, final stages of SI&T may actually be accomplished at the user site prior to system testing.

Acceptance Testing

Acceptance testing can take on a range of implementations. In one-of-a-kind systems, there is typically a single customer. When the developing organization is satisfied that the entire system operates in a manner that meets all functional requirements, it is then ready to assure the customer that this is indeed the case. The purpose of the acceptance test plans and procedures is to establish a well-defined previously agreed-to point at which the system can be transferred to the customer. Successful completion of acceptance testing and the subsequent signing off of that completion define the point at which the implementation is officially completed—the same point at which operations and maintenance contracts usually begin. It is very important to seek early agreement between the vendor and the customer on the philosophy of how the system will be accepted and the details of actual acceptance testing. These test plans and procedures for acceptance testing need to be known by the PDT leader early on so that adequate planning can take place in the design of system integration and system-level testing prior to acceptance testing. The overall aim is to avoid surprises and misunderstandings during acceptance testing.

In some cases, acceptance testing is merely a repeat of the implementing organizations' system-level tests with the customer participating and/or observing. In simpler systems, the customer may execute an entire repeat of system-level testing with the implementors' support.

In complex systems, other realistic strategies may be required. In many cases, the user(s) does not have the resources to conduct complete system-level testing, or the system may be so complex that testing every possible logical combination is prohibitive. In these cases, acceptance tests typically amount to a statistical validation of system capability. For example, consider an actual case of a software system designed to carry out automated test of inertial navigation computers at the board level upon the completion of manufacture. It was agreed that the acceptance test for the fault isolation software would consist of a number of software test runs. Prior to each run, a tester from the customer organization would remove a chip from a board known to be good, bend one of the pins upward, and replace the chip back in its socket so that the affected pin did not make contact. The software under test would then be run to see if the fault was successfully isolated to that particular chip. It was further agreed that carrying out this procedure for every pin on every chip on every board would require an excessive amount of time. The customer agreed that, if its personnel could carry out this procedure randomly eight hours a day for a period of two weeks without encountering an error in the fault isolation software, this would constitute grounds for formal system acceptance. This statistical test was carried out successfully and the system was accepted after the first two weeks of testing. The important point is that the approach was negotiated early, before the product design review, and this approach impacted the test design throughout implementation.

Measurement of performance parameters such as response times, error rates, propagation coverage, acceleration, and other related functional requirements are also statistical measurements requiring negotiated, realizable methodologies. Simulation of external stimuli to the system are often required. Measurement of system availability, another statistic, is often accomplished over an agreed-to period in an operational setting. A typical approach to availability testing is to reset the test period clock upon correction of each failure that falls short of the desired mean time to failure figure. This approach is favorable to the customer. Limits to the number of resets to be tolerated must be worked out in the interest of both parties.

Details of acceptance test procedures are often overlooked until late in the implementation process, perhaps because it seems so far away at the beginning of an intricate and lengthy project. Be aware of the importance of addressing this issue with the design of sound procedures in order to avoid unforeseen changes and attending schedule alterations. This is an area of great importance to the developing organization since it defines the criteria for achievement of the end point of the system development effort.

Adequate testing of commercial systems designed for use by a large number of customers presents an additional problem. One approach in this case is the use of alpha and beta testing concepts. In alpha testing, the product is taken to a specific representative customer where testing takes place as a joint effort between the developer and the customer. Beta testing involves the distribution of the product to a number of prospective customers who each test the product in their own operational settings. A combination of the two strategies is often used where the alpha test is designed to reveal the majority of customer recommendations and the beta test is then used to refine these findings over a larger sample base.

The design of all test plans and procedures is closely integrated with all aspects of the detailed hardware and software system design.

DONE

The successful execution of Phase C has allowed us to get to market faster, on schedule, within cost and with a quality product that truly meets user needs. We have done this through the use of the most up-to-date, modern systems engineering techniques. Among these has been the use of a well-chosen, highly motivated concurrent product development team (PDT). The efficiency of the team was greatly enhanced through co-location. The teams' work was thorough and complete through the construction and use of a sound work breakdown structure (WBS) that accurately reflected the actual work to be done. The WBS was used along with product priorities for the structured decomposition of requirements and design trade-off analysis. Interface design has been greatly facilitated through the methodic iteration of N^2 diagrams. Our reviews were well planned and productive, and our documentation is complete and traceable to top-level functional requirements. And all of this happened without reorganization because management empowered the product development team to make decisions and succeed. It has been at once difficult, and incredibly easy.

Something else happened. Everyone wanted to come to work, and everyone is proud of the outcome. They all want to do it again.

ENDNOTES

1. R. Lano, "The N^2 Chart," *TRW SS-77-04* (One Space Park, Redondo Beach, CA: TRW, Inc., Nov., 1977.

2. H. Zimmerman, "OSI Reference Model—The ISO Model of Architecture for Open System Interconnection," *IEEE Transactions on Communications,* Vol. Com-28, Number 4, April 1980.

Part 7

Getting Started

There is no question that we can always do better. Hopefully this book has provided some ideas on *how* to do better. The question is, how do we begin? For most of us that means we need to constructively approach our management for support. Actually, managers are better these days than they have been in previous decades. In our favor is the fact that more and more managers today are engaging in self-examination and abandoning autocratic attitudes. Perspectives have changed, and continue to change, mostly due to the many positive aspects of the quality movement. Self-awareness has been heightened, in particular, through the increasing number of organizations seeking to attain certification. This is good, because many organizations are getting in the spirit of looking at themselves. If you want to get the ball rolling toward full-scale team based product development, here are some thoughts on getting started. They have worked for me.

Before you talk to anyone, do some thinking on your own. Get yourself prepared by making a plan. A starting point is to write down the differences between the way product development is currently organized and carried out in your organization and the guidelines you may now wish to incorporate, having read this book. The following are questions to ask yourself for each phase. A checklist of questions for the Product Development Feasibility Phase (Phase A) would include

A1. When you consider user needs for a new or improved product, do you also consider user constraints for product use and user resources for the use and maintenance of the product? Should additional primary or secondary market research be a part of your feasibility budget and schedule? Do you consider needs, constraints, and resources of your own organization, any

287

partners, suppliers, and sponsors? Do you have your own organization's support? Are lines of communication, responsibility, and authority agreed upon with partners, suppliers, and sponsors?

A2. Do you routinely develop a mission statement for each new product or product improvement?

A3. Do you develop top-level achievable design concepts to support feasibility?

A4. Do you consider inherited equipment to support the design concepts?

A5. Do you conduct top-level risk analysis to support feasibility?

A6. Do you develop top-level schedule and cost estimates based on parametrics, analogy, or other experience?

A7. Do you think about, understand, and recommend a proper product development paradigm up front?

A8. Do you develop a first cut at a return on investment analysis?

A9. Do you use a formal team consisting of appropriate corporate-wide expertise to assure that all bases have been covered during the feasibility study? (Refer to "Who's on the Team?" Part 4). Are the team members co-located? Does the team itself have a budget, schedule, and agreed-upon deliverables? Does the team leader faithfully conduct *weekly* status meetings, efficiently direct activities, and assign action items?

A10. Does the team leader, with team support, conduct a formal review for peers and management on study findings, with recommendations at the conclusion of the team's work?

A checklist for the Detailed Product Development Planning Phase (Phase B) would include

B1. Do you begin planning by refining and finalizing user needs? Are you focused on the right user community? Do you understand the user environment and culture—local, regional, or national? Should further market research support detailed planning?

B2. Do you develop a specific work breakdown structure (WBS) for each new or improved product? If you adopt an existing WBS routinely used by your organization, does it fit *all* of your needs? Should it be modified to reflect the actual work you plan to do?

B3. Does your planned organization for product development map onto the work to be done? Are you faithfully staffing the content of your WBS or are you organizing to do something else?

B4. Do you derive product development priorities from the product mission statement? Do you plan to use these priorities to support complete analysis and consistent decision making throughout the actual execution of product development? When you develop a product, does *everyone* understand your mission and your priorities, or does every department, and everyone in every department, have a different idea about what is going on?

B5. Do you plan for review and reporting structures? Are those structures consistent with the WBS with regard to content and who reports to whom? Is the content of your reports complete and is the form and format consistent, or does everybody do his or her own thing?

B6. Do you identify items for technical margin management in advance? Do you develop strategies in advance for monitoring and controlling these?

B7. Do you develop plans for documentation in support of development, production, dealers, maintenance, and in direct support to the users? Is your documentation form and format integrated and consistent? Are sentences or sections repeated in different documents, making configuration management more difficult? Do you use the WBS to plan for who is going to produce what documents? Do you plan for who is going to update and control the accuracy of documents? Do you make plans for ease in document access?

B8. Do you devote enough time to identifying major risk items and devise strategies to control their effects? Do you identify specific times in your schedules to implement alternative strategies or controls?

B9. Do you plan for a configuration management (CM) approach that will support your schedule? Is your product development team empowered to manage CM by itself? If not, and you adopt existing CM mechanisms, are they the right ones for your specific product development undertaking? Should you plan to revise them, or even build your own?

B10. Do you have supplier agreements in place? Do they understand what they are supposed to deliver, when they are supposed to deliver it, and your quality goals? Do they understand your total mission? Are they a part of your team?

B11. Do you construct a complete resource plan, including integrated costing and scheduling? Do you use experts to do this in a complete and consistent manner?

B12. Do you develop a complete Product Development Team Management Plan?

B13. Do you develop a detailed return on investment analysis as a part of your planning?

B14. Do you use a formal team consisting of appropriate corporate-wide expertise to ensure all bases have been covered during detailed product development planning? (Refer to "Who's on the team," Part 5). Are the team members co-located? Does the team itself have a budget, schedule, and agreed-upon deliverables? Does the team leader faithfully conduct *weekly* status meetings, direct activities efficiently, and assign action items?

B15. Does the team leader, with team support, conduct a formal review for peers and management on study findings with recommendations at the conclusion of the team's work?

A checklist of questions for the actual Concurrent Product Development Phase (Phase C), include

C1. Do you begin by finalizing and staffing the WBS, including the formation of the product development team?

C2. Do you proceed by immediately finalizing the product development team management plan?

C3. Do you next finalize user needs and develop formal product functional requirements for signature?

C4. Do you use product priorities together with the product breakdown structure in matrix form to develop requirements flowdown to subsystem levels, resulting in formal product level and subsystem specifications?

C5. Do you finalize your CM plan? Is it supportive of the work you need to do?

C6. Do you use the N^2 diagram approach for the top-down definition of product interfaces to ensure completeness and detailed understanding? If not, what is the alternative?

C7. Do you use product priorities with the product breakdown structure in matrix form in order to provide completeness in identifying issues to be addressed in trade-off analysis as well as a guide for decision making?

C8. Do you develop product and lower-level test requirements based on functional requirements?

C9. Do you develop a complete product development logistics support plan that includes supply support, equipment for design and prototyping, test equipment and support, transportation and handling, facilities, personnel and training, maintenance and technical documentation handling?

C10. Do you carry out sufficient testing as required for parts, assemblies, subsystems, product integration, and at the product level for development and production? Does your design have enough test points?

C11. Do you use a formal product development team (PDT), consisting of appropriate corporate-wide expertise to ensure that all bases have been covered during the development phase? (See "Who's on the team?," Part 6). Are the team members co-located? Do you have sufficient user representation on the team? Does the team leader faithfully conduct *weekly* status meetings, direct activities efficiently, and assign action items? Does the PDT stick to sound systems engineering practices, as outlined in this book?

C12. Do you conduct sufficient formal reviews that include, at a minimum, a product requirements review, a preliminary design review, and a critical design review?

It is helpful at first to consider application of these guidelines in the context of a candidate pilot product. Pick a simple project that is familiar to everyone. By selecting a pilot project, you are not asking the whole organization to adapt to radical change. You are simply suggesting a controlled and limited test on a specific product. It may also be helpful at this point to have in mind specific candidate team leaders for each phase. The leaders should be people known to be both innovative and knowledgeable, and who possess outstanding skills in human relations. At this point, these are not final selections, just examples to use in the presentation of your ideas.

Think about what you might like to present to the appropriate manager(s) in your organization. For example, a 50-minute presentation could involve the following slide material:

1. Define team based product development as: a process that utilizes corporate-wide interdisciplinary expertise for: (a) initial assessment of product development feasibility, (b) complete planning for product development, and (c) concurrent execution of product development.

2. List what, if anything, is being neglected by omission in your current processes. Do this by comparing your current processes with the guidelines noted above.

3. Give examples of team players by title. (Review Parts 4, 5, and 6 for team membership examples.) Tailor these to your organization.

4. List the advantages of team based product development. (See the points made in the Introduction, and add your own thoughts.)

5. List some successful examples. (See Part 2. Add other examples of which you may be aware.)

6. Use a few slides to make a proposal. Ask for permission to carry out a feasibility task. Provide a budget and a schedule for the task. List your deliverables to include identification of a pilot product, a candidate pilot manager, and potential team memberships by title. Include perceived measurable advantages and recommendations. Propose to report back in X weeks (less than a month) with your findings. Your findings will also include a plan to approach appropriate managers, enlisting their help in making specific preliminary assignments. You are not making assignments yet—just a plan on how to enlist the right people. You are going step by step. Tell them you are not trying to be a knight on a white charger. You are only suggesting that more concurrent input up front has proven to save time and improve product fidelity and quality. None of this is threatening. In fact, it offers interesting promise. When you report back, you will be asking for their ideas for improving your plan and, of course, approval to go ahead.

Once you have your ideas together, and you feel comfortable with your ability to address all questions that might come up, then (and only then) are you ready to talk about it. Feel free to pass this book around. The first person to talk with is your immediate supervisor. Ask what he or she thinks about it and try to enlist support and involvement. Ask for suggestions for improvement of your plan. Summarize the ideas and procedures from this book that you want to use. Add your own well-tailored stuff, consistent with the way you do business and in a form that makes sense. Stress that you are not trying to change the existing organization but to better utilize its existing resources and potential capabilities. Propose that together you prepare, and ask for, a presentation to management that is high enough up on the corporate ladder to make it all happen.

There are at least two common rebuttals you should be prepared for. There was a time when I was in marketing. I had some pretty good products. Still, over time, I couldn't help notice that when potential customers really didn't want to buy, they seemed to come up with rebuttals that followed similar lines. The exact words, of course, were a little different each time, but it seemed, more often than not, that I was hearing the same things over again. In fact, it was not difficult to boil it down to two distinct rebuttals. From then on, I carried a little note in my right-hand suit pocket and another note in my left-hand suit pocket. When I heard either of the standard rebuttals, I would pull the appropriate response out and show it to the prospect.

One of the notes said, "That's a good idea, but it won't work here."

The other one said, "We're already doing that."

Of course, I was young back then and possibly a bit of a wise guy. Certainly my little trick did not always result in deep introspection on the part of my prospects. Now, as you may have noticed, I am somewhat more diplomatic.

Having read this book, you now have the ammunition to truly understand that neither of these objections is really valid in any product development setting.

First of all, team based product development *will* work. It will work anywhere. This is because, if you do it right

1. You're going to get to market faster.
2. Your product is going to do what the customer wants it to do.
3. Your product is going to be one of quality.
4. Your product is going to be right for your organization.
5. The probability of errors of omission is going to be significantly reduced.
6. You will go through the development process on time and within budget because you have planned properly.
7. Your team will go through a very positive experience involving the building of individuals' self-esteem and pride in ownership.
8. Everyone involved will want to come to work.

When people say, "We're already doing that," they are most often referring to some other kind of teams. What people really mean is that they have "teams" somewhere. That is, they use the term. Typically, they are teams that meet on no particular schedule, or "as soon as there is a problem." Or, they are teams at lower levels of product decomposition focusing solely on the technical detail. The teams we are talking about exist at the top level of each product development phase. They stay together through the completion of their phase. They are completely multidisciplinary. They address *all* the issues. They meet every week, like clockwork, with well-planned intermediate products, reviews, schedules, and goals.

In your presentation to management, you may want to use one or two of your slides, or backup slides, to address these two common objections when raised. But remain positive. If it is said, "We're already doing that," simply reply "That's very true, we are using teams, but there is a little different slant to these teams. . . ."

Report back in a few weeks, within a month. Keep it fresh in everyone's mind. Pick a new product development, or product improvement effort, that is likely to succeed. If you haven't worked with teams before, there will be enough to go through the first time out without picking a loser project at the outset. Suggest that the pilot development effort you have selected is one from which we can all learn valuable lessons, enjoy a high probability of success, and set a positive example for future work. List the specific departments or divisions within the organization that you propose to approach. Ask advice from your audience at the end of your presentation. Ask them for ideas on how to do it better. But stick to the fundamentals. Take charge of the discussion and lead it.

There is an alternative approach. But there is a condition. If you happen to already have the responsibility for product development at the model level, and you also have the authority to lead that development (i.e., management believes in you), then you can form a concurrent product development team and execute at least Phase C within your own authority. I have done this a number of times. Simply visit the immediate managers, or supervisors, of the people you want on your team, enlist their support, and then get the job done successfully. I know this can be done because I have personally formed successful product development teams within existing organizational structures. Significantly, this was accomplished on a number of occasions where development efforts had previously failed. *Each failure was directly traceable to the lack of one or more of the guidelines and tenets covered in this book.* Each success involved the complete team approach with the right people, the right paradigm, a good set of priorities, constant user interaction, plenty of listening, and the all-important delegation of authority to team leaders to lead and act. There is no question that team leaders must have authority for teams to work successfully. Authority is readily delegated by good managers, that is, leaders. *Do not accept responsibility without authority.*

There is a distinct advantage to this alternate approach. Rather than just talking about what should or shouldn't be done, you are simply doing it. There is nothing like success to gain attention and acceptance of new ideas. It works better than talking about it. In my experience, once you have gone through the complete Phase C product development process, and succeeded where others did not do as well, upper management suddenly wants to know what you did. They also become very open to the idea of expanding the effective use of teams to phases A and B. In other words, you succeed by doing.

In any case, you are well positioned. You are armed with inherent good sense, a good understanding of what should be done and in what sequence, and good benchmarked experience to draw on. And don't ever forget one more very important concept. With team based product development, you are poised to employ the single most powerful tool that any of us can marshal— a concept that's 10,000 years old. That tool is the latent need in everyone for individual worth, pride, responsibility, freedom from fear, and dignity in the workplace. It is what makes teams work. It is what makes quality work. This immensely powerful tool silently awaits you. Take it up! Use it!

Glossary
of Acronyms

ACWP	Actual cost of work performed
AI	Action item
AR	Anomaly report
ASQ	American Society for Quality
BCWP	Budgeted cost of work performed
BCWS	Budgeted cost of work scheduled
BIT	Built-in-test
BITE	Built-in-test equipment
CCB	Configuration control board
CDR	Critical design review
CE	Customer engineer
CEP	Circular error probability
CI	Configuration item
CM	Configuration management
CMP	Configuration management plan
COGE	Cognizant engineer
CPC	Computer program component
CPCI	Computer program configuration item
CPM	Critical path method
CPS	Critical path scheduling
CPU	Central processing unit
CRT	Cathode ray tube

DBM	Data base management
DCR	Delivery commitment review
DIV	Division
DLL	Design language
ECP	Engineering change proposal
ECR	Engineering change request
ECU	Electrical control unit
EMI	Electromagnetic interference
FR	Functional requirement (see **FRP**)
FRD	Functional requirements document
FRP	Failure report (also commonly referred to as an FR)
GPSS	General purpose simulation system
HP	Horsepower
HVAC	Heating, ventilation, air conditioning
HWCI	Hardware configuration item
ICD	Interface control document
ILS	Integrated logistics support
IOC	Initial operational capability
IEEE	Institute of electrical and electronic engineers
INCOSE	International Council on Systems Engineering
I&T	Integration and test
JIT	Just-in-time
LRU	Line repaceable unit
MDT	Mean down time
MPBS	Mission product breakdown structure
MPG	Miles per gallon
MTBF	Mean time between failures
MTTR	Mean time to restore
OOP	Object-oriented programming
OSI	Open system interconnect
PDF	Program development folder
PDL	Program design language
PDR	Preliminary design review
PDT	Product design team

PERT	Project evaluation and review technique
P/F	Pass / fail
PPAR	Production parts approval review
PRR	Product requirements review
PSR	Pre-ship review
QC	Quality control
QFD	Quality function deployment
RAM	Reliability, availability, maintainability
RDM	Rapid development method
RFP	Request for proposal
RFQ	Request for quotation
RJE	Remote job entry
ROI	Return on investment
SDD	Software design document
SE	Systems engineering
SI&T	System integration & test
SMP	System maintenance plan
SOA	State of the art
SPC	Statistical process control
SRD	Software requirements document
SRR	Systems requirements review
SSD	Software specifications document
S/S	Subsystem
SWCI	Software configuration item
T&H	Transportation and handling
TP&P	Test plans & procedures
TRR	Test readiness review
TTO	Transfer to operations
VCR	Video cassette recorder
VLSI	Very large scale integration
V&V	Verification and validation
WBS	Work breakdown structure

INDEX

A

Acceptance testing, 283-285
Accountability, 8, 11
Accounting, 176, 193
Acronym glossary, 297-299
Action items, 28, 31
Actual cost of work performed
  (ACWP), 158
Ahmadabad, India (example of
  teams), 13
Allocated baseline, 178
  staircase paradigm, 225
Alternative approach to
  implementation of product
  development phases, 294-295
American Productivity and Quality
  Center, 12
Anger, 17
Anomaly reports (ARs), 184
Attitudes, 10
Attribute utility trade curves, 240-241
Auditing, 175, 190-193
Authority and responsibility, 21

B

Baselines, 176-178
BCWP. See Budgeted cost of work
  performed
BCWS. See Budgeted cost of work
  scheduled
BIT. See Built-in-test
BITE. See Built-in test equipment
"Black hole" of software
  development, 282

Block diagrams, 88, 111
Bosses
  definition of, 10
  versus leaders, 16, 26
Brainstorming, 11
Brassboarding, 228, 266, 267
Breadboarding, 228, 265-267
British mining (example of teams),
  12-13
Budgeted cost of work performed
  (BCWP), 158
Budgeted cost of work scheduled
  (BCWS), 158
Budgeting resources, 163
"Build It, Test It, Fix It," 56
Built-in test equipment (BITE), 274
Built-in-test (BIT), 274-276
Buzz words, 35

C

Cadillac (example of teams), 14-16
CCB. See Change control board
CDR. See Critical design review
CEP. See Circular error probability
CEs. See Customer engineers
Change, 183
Change control board (CCB), 63,
  179-180
Circular error probability (CEP),
  240
Civility, 28
CMP. See Configuration
  management plan
Co-locations, 29